BUILDING
PROOFS

A Practical Guide

D1248394

BUILDING PROOFS

A Practical Guide

Suely Oliveira
David Stewart
The University of Iowa, USA

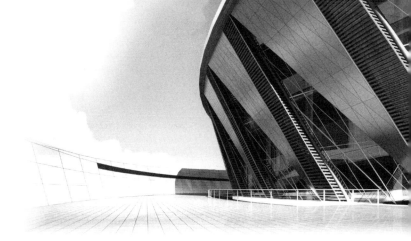

 **World Scientific**

NEW JERSEY · LONDON · SINGAPORE · BEIJING · SHANGHAI · HONG KONG · TAIPEI · CHENNAI

Published by

World Scientific Publishing Co. Pte. Ltd.

5 Toh Tuck Link, Singapore 596224

USA office: 27 Warren Street, Suite 401-402, Hackensack, NJ 07601

UK office: 57 Shelton Street, Covent Garden, London WC2H 9HE

Library of Congress Cataloging-in-Publication Data
Oliveira, Suely.
  Building proofs : a practical guide / by Suely Oliveira (The University of Iowa, USA),
David Stewart (The University of Iowa, USA).
    pages cm
  Includes bibliographical references and index.
  ISBN 978-9814641296 (hardcover : alk. paper) -- ISBN 978-9814641302 (pbk. : alk. paper)
  1. Proof theory. 2. Logic, Symbolic and mathematical. 3. Mathematics. I. Stewart, David,
1961–   II. Title.
  QA9.54.O46 2015
  511.3'6--dc23

                                                    2015012343

British Library Cataloguing-in-Publication Data
A catalogue record for this book is available from the British Library.

Printed in Singapore

Preface

This is a small book to guide mathematics and computer science students in the art of proofs and proof-making. Examples are taken from many different parts of mathematics (number theory, calculus and analysis, geometry, algebra, and algorithms). Most of the material assumes that students have taken the standard courses on calculus and linear algebra before going on to courses called "analysis" or "abstract algebra" or something similar. The focus is less on the specific content, than ideas on how to create proofs appropriate for whatever problem you face. The level of the material ranges from advice for the complete beginner, to ideas for handling the development of new theories. Whatever the level, this is meant to be a book to keep around and pull out when trying to puzzle out a new proof.

There are other books about learning to prove mathematical statements. Most of them are written as textbooks for some course about fundamentals of mathematics or logic or how to prove things. This is not: writing proofs should not be put off until after you learn "logic" or "sets". By the time you have taken calculus and linear algebra, you already know quite a lot of mathematics. Jump in! Use what you already know. Then add to your knowledge. It is not as scary or as difficult as some make it seem.

A note about the references: There are a few references to textbooks given in this book. These are not necessary reading, but are included as supplements. Some, such as Russell and Whiteheads' *Principia Mathematica* and Frege's *Begriffschrift*, are included as historical references, and not easy to read. If you are wanting to extend your knowledge or apply to a particular course or area, pick up a textbook for that course or area. Don't be afraid to look around for other books to help your understanding. But be aware that different books have different emphases and notation. What you should look for is clarity of explanation. While poets often try for mystery, mathematicians praise clarity and correctness above all.

   A note about the exercises: It is essentially impossible to learn mathematics without doing mathematics. For this reason there are a number of exercises at the end of each chapter. These range in topic and difficulty. Particularly difficult ones are marked by ⚠. More difficult *sections* are marked by an asterisk (*), and can be skipped on a first reading.

Suely Oliveira and David Stewart
Iowa City, Iowa

Contents

	2.1	Propositional calculus	32
		2.1.1	Propositional logic and truth tables	32
		2.1.2	Precedence	33
		2.1.3	Truth tables	34
		2.1.4	Tautologies	34
		2.1.5	Rules of inference	35
		2.1.6	Implication via assumption*	36
		2.1.7	Inference steps	36
		2.1.8	The algebra of propositions	37
		2.1.9	Boolean algebra	38
	2.2	Expressions, predicates, and quantifiers	39
	2.3	Rules of inference	41
		2.3.1	Rules of inference for propositional calculus	41
		2.3.2	Rules of inference with quantifiers	41
	2.4	Axioms of equality and inequality	42
	2.5	Dealing with sets	43
		2.5.1	Set operations	44
		2.5.2	Special sets	45
		2.5.3	Functions	45
	2.6	Proof by induction	46
		2.6.1	Some sums	48
		2.6.2	Fibonacci numbers	49
		2.6.3	Egyptian fractions	50
	2.7	Proofs and algorithms	52
		2.7.1	Correctness by design*	56
	2.8	Exercises	59

3.	Discrete and continuous	63

	3.1	Inequalities	63
		3.1.1	Some classic inequalities	64
		3.1.2	Convex functions	66
	3.2	Some proofs in number theory	69
		3.2.1	Division algorithm and gcd	69
		3.2.2	Prime numbers	73
		3.2.3	"There are infinitely many primes"	74
	3.3	Calculate the same thing in two different ways	75
		3.3.1	Sums of degrees in graphs	76

Chapter 1

Getting started

There comes a point for any student of mathematics and computer science where you are asked to "show that" or "prove that" something is true.

At first, it is like stage fright: you don't know where to start. But once you begin the process, it can flow remarkably smoothly. Beginners can find some of the moves a little strange at first ("proof by contradiction", for example), but with familiarity, the ideas behind the moves becomes clear. Others are easier to follow (such as "unwrap the definition") that can soon become automatic.

1.1 A first example

Let's get started with a famous result that (according to legend) cost the discoverer his life. We will go through the proof step by step. Don't worry if some steps involve some knowledge that you were not aware of. Look for the structure of the proof: how the parts work together.

Theorem 1.1. $\sqrt{2}$ *is irrational.*

Proof. Suppose that $\sqrt{2}$ is *not* irrational. That is, suppose $\sqrt{2}$ is actually rational.   *1*

> *This uses proof by contradiction. We start by supposing the claim is false, and then show that this leads to an impossibility. This is often a good approach for proving negative results, showing that something* does not *happen.*

Then we can write $\sqrt{2} = a/b$ where a and b are integers with $b \neq 0$.   *2*

> *This unwraps the definition of being rational.*

Assume without loss of generality (AWLOG) that there are no common factors of   *3*

a and b.

> *If there were any common factors we could cancel them out, and repeatedly do this until there are no common factors left.*

Then $2 = (\sqrt{2})^2 = (a/b)^2 = a^2/b^2$. **4**

> *This unwraps the definition of "$\sqrt{2}$", plus does a little bit of obvious calculation.*
> *There are no obvious rules about fractions to use here. But we can reduce the question to one about whole numbers.*

Multiply both sides by b^2. This gives $2b^2 = a^2$. **5**

> *That's better! Now we can use some knowledge about even and odd numbers.*

Since $2b^2$ is even, we must have a^2 even. Now a is either even or odd. **6**

> *Now we have two cases to consider. Let's try them in turn. We want to show that neither case is possible.*
> *The first step is to unwrap the definition of "odd".*

If a is odd, then $a = 2c + 1$ for some integer c. Then $a^2 = (2c + 1)^2 = 4c^2 + 4c + 1 = 2(2c^2 + 2c) + 1$ is also odd, which contradicts the fact that a^2 is even. **7**

> *Next we have the case where a is even.*

If a is even, then $a = 2c$ for some integer c. **8**

> *We just unwrapped the definition of "even".*

Therefore $a^2 = (2c)^2 = 4c^2 = 2b^2$. Dividing by 2 gives $b^2 = 2c^2$. Using the previous argument, b must be even as well. **9**

> *Previously we had $a^2 = 2b^2$, and showed that a must be even. Now $b^2 = 2c^2$ so b must be even.*

Therefore, both a and b are even and so both are divisible by two. This means that two is a common factor, which contradicts our assumption that a and b have no common factors. **10**

> *This contradicts the "AWLOG" assumption above. Since we can always ensure that this assumption holds if $\sqrt{2}$ is rational, we have indeed found a contradiction.*

Thus a cannot be either even or odd, which is impossible for a whole number. *11*
Therefore our original assumption must be false.

> *It is always a good idea to make sure that the statement of the*
> *theorem is indeed proved. Don't leave the reader "hanging in*
> *the air", unsure that the proof really does the job.*

That is, $\sqrt{2}$ is an irrational number. $\square$ *12*

Let's review the kinds of moves that were used in this proof:

- *proof by contradiction*: (line *1*)
  This is often known by the Latin phrase *reductio ad absurdam* (reduction
  to absurdity). Proofs by contradiction are often called *indirect proofs*;
  other proofs are called direct proofs.
  Since *"irrational"* means *"not rational"*, we are really being asked to
  prove a negative statement. In such a case, proof by contradiction is often the best approach. Otherwise, direct proofs tend to be better.

- *unwrap the definition*: (lines *1, 2, 4*)
  Unwrapping definitions is a basic move in creating proofs, and very often
  the first step in a proof. Often we unwrap a well-known definition (such
  as the definition of *"irrational"* or *"$\sqrt{2}$"*), or perhaps a definition of a
  variable you created earlier in the proof.
  At the very beginning of the proof, we face the task *"Show that $\sqrt{2}$ is*
  irrational". The next question we should ask ourselves is *"What does it*
  mean for $\sqrt{2}$ to be irrational?" This means unwrapping the definition of
  "irrational": it means *"not rational"*. This is used in line *1*. After using
  proof by contradiction, we have to ask ourselves *"What does it mean for*
  a number to be rational?" It means *"can be written as a fraction a/b of*
  whole numbers with $b \neq 0$". This is used in line *2*. Finally, what is $\sqrt{2}$?
  It is the positive real number where $(\sqrt{2})^2 = 2$. This is not an incidental
  fact about $\sqrt{2}$. That is its definition. It is also used in line *4*.

- *reduction to a simpler kind of problem*: (line *5*)
  Mathematicians constantly seek to reduce complex problems and tasks
  to simpler problems and tasks. The truth is that most of us do not know
  much about irrational numbers; we know more about rational numbers,
  but whole numbers are much easier to pin down. In particular, whole
  numbers are either even or odd, which becomes the crux of the proof.

- *treating cases separately*: (lines *6*, *7*, *8*)

  In line 6 we come to the statement that "*a is either even or odd*". We do not yet know which case is true, but we know that *one* of them is true. We then proceed to look at both possibilities. In other proofs we may need to consider many possible alternatives. Some proofs in the area of graph theory and combinatorics involve little else than identifying separate cases and treating them. It is best to have at most binary choices (either *this case* or *that case*) at any step of a proof. Cases can be nested (so *case 1* can be divided into *case 1a* and *case 1b*, and then into *case 1a(i)* and *case 1a(ii)*, etc.). Going too far with this makes for complicated and confusing proofs. It is better to have just a few cases and keep the flow of the proof as much as possible without breaking it up into small branches and twigs.

- *assumptions without loss of generality*: (line *3*)

  Making an *assumption without loss of generality* (*AWLOG*) must be done carefully. You need to check that the assumption made really can be made without eliminating certain possibilities. In this case, when we write a number as a fraction a/b, $b \neq 0$ it is always possible to ensure that a and b have no common factors. We can even assume without loss of generality that b is positive. (We might have to change the sign of a if b is negative.) We can assume without loss of generality that a is positive. (We might have to change the sign of b if a is negative.) But we cannot assume without loss of generality that *both* a and b are positive.

  Making a false *assumption without loss of generality* is a trap that even professional mathematicians can fall into. Where possible, convert an *assumption without loss of generality* into separate cases (such as case 1: $b > 0$ and case 2: $b < 0$) and deal with each case separately.

- *say what you have done*: (line *12*)

  Saying what you have just proved is very helpful for the reader, so that it is clear that this is the end of the proof. The closing statement should match, in part, the statement of the theorem.

1.2   The starting line: definitions and axioms

In order to prove anything we need definitions and axioms. Definitions reduce more complex concepts into simpler concepts. Axioms are basic facts or assumptions that we take to be true without needing further explanation or justification.

1.2.1 *Definitions*

Usually the first step we need to use in a proof is to *unwrap the definition(s)*. We need to ask "*What do the concepts in the theorem mean?*" In the example above, we want to ask what it means for $\sqrt{2}$ to be irrational. It means that $\sqrt{2}$ is not rational. But what does that mean? It means that $\sqrt{2}$ is not equal to any a/b where a and b are whole numbers with $b \neq 0$. We also need to ask, "*What is $\sqrt{2}$?*" (It is the positive number x where $x^2 = 2$.)

We could keep going like this. ("*What is 2?*") But there is usually a place where it becomes obvious that there is no reason to continue breaking down concepts, at least for the purposes of creating a proof. For example, consider the definition of "limit":[1]

Definition 1.1. *The statement* "$\lim_{x \to a} f(x) = L$" *means that for any $\epsilon > 0$ there is a $\delta > 0$ where $0 < |x - a| < \delta$ implies $|f(x) - L| < \epsilon$; this is also written* "$f(x) \to L$ *as* $x \to a$."

From this definition , we can prove a number of rules about limits: Assuming $\lim_{x \to a} f(x)$ and $\lim_{x \to a} g(x)$ exist,

$$\lim_{x \to a} f(x) + g(x) = \lim_{x \to a} f(x) + \lim_{x \to a} g(x), \tag{1.1}$$

$$\lim_{x \to a} f(x) - g(x) = \lim_{x \to a} f(x) - \lim_{x \to a} g(x), \tag{1.2}$$

$$\lim_{x \to a} f(x) \cdot g(x) = \lim_{x \to a} f(x) \cdot \lim_{x \to a} g(x), \tag{1.3}$$

$$\lim_{x \to a} \frac{f(x)}{g(x)} = \frac{\lim_{x \to a} f(x)}{\lim_{x \to a} g(x)}, \tag{1.4}$$

$$\text{provided } \lim_{x \to a} g(x) \neq 0.$$

For most problems we rely on these rules. But sometimes, we need to go back to the definition. For example, if we want to properly prove the squeeze theorem below, then we will need to go back to the definition.

Theorem 1.2 (Squeeze theorem). *If* $f(x) \leq h(x) \leq g(x)$ *for all* x *and* $\lim_{x \to a} f(x) = \lim_{x \to a} g(x) = L$, *then* $\lim_{x \to a} h(x) = L$.

---

[1]This definition is probably the most difficult definition that is encountered in calculus courses. It usually takes several courses to properly understand and use this definition.

1.2.2 *Axioms*

Axioms are our basic assumptions. These assumptions do not require any explanation or justification, at least in a proof. The choice of axioms may be a matter for discussion, but once these are chosen they can be used without further questions.

Starting calculus, most students are told what a limit is in vague terms, often with a picture to emphasize the geometric idea of continuity. Then a number of rules are given, like the rules (1.1–1.4) above. In this situation, the rules can be treated as axioms.

Later, the limit is defined in terms of inequalities (Definition 1.1), and we use the properties of real numbers and arithmetic operations as the axioms. Later still, we might start with Peano's axioms for the natural numbers $\mathbb{N}$, and from that define the real numbers $\mathbb{R}$ and their properties. Even later, we might start with the axioms of set theory, and use this to define the natural numbers.

Exactly what our axioms should be depends on our starting point and our aims. For a first course in calculus, the limit rules can be taken as axioms. For a starting course on mathematical analysis, the properties of real numbers can be taken as the axioms; the limit rules are then theorems to prove. An introductory course on the foundations of mathematics might start with Peano's axioms and define the real numbers; a second course on the foundations of mathematics might start from the axioms of set theory and define $\mathbb{N}$ so that Peano axioms become theorems about $\mathbb{N}$.

In abstract algebra, the axioms can be part of the definitions. For example:

Definition 1.2. *A* group *is a set G with a binary operation $*$, a unary operation $(\cdot)^{-1}$, and a special element e with the following properties:*

$$(x * y) * z = x * (y * z) \tag{1.5}$$

$$x * e = e * x = x, \tag{1.6}$$

$$x * x^{-1} = x^{-1} * x = e \tag{1.7}$$

for all x, y and z in G.

The properties listed become the axioms for a group.

If you are unsure about what axioms to use, check your textbook and see what axioms are used in it. Ask your instructor. Sometimes the question asked will say *"Assuming that ..."* which makes clear any additional axioms you might need.

1.3 Matching and dummy variables

Using a formula in a strange context can be difficult. It is important to be able to match a general formula with our own current situation. This is not usually difficult, but practice makes perfect!

1.3.1 *Matching up expressions*

Let's get some practice at matching expressions.

1.3.1.1 *A calculus example*

In a calculus course you might be asked to compute the derivative

$$\frac{d}{dt}(t \sin t).$$

Your instructor might mention the product rule. After all, $t \sin t$ is the product of t and $\sin t$. What does the product rule look like?

$$\frac{d}{dx}(u(x)\,v(x)) = \frac{du}{dx}(x)\,v(x) + u(x)\,\frac{dv}{dx}(x).$$

Clearly we need to match $\frac{d}{dt}(t \sin t)$ with $\frac{d}{dx}(u(x)\,v(x))$. The variable t in $\frac{d}{dt}(t \sin t)$ matches the variable x in $\frac{d}{dx}(u(x)\,v(x))$; then $u(x)$ can match t and $v(x)$ can match $\sin t$. Then

$$\frac{d}{dt}(t \sin t) = \frac{dt}{dt}\sin t + t\,\frac{d\sin t}{dt}$$
$$= 1\,\sin t + t\,\cos t = \sin t + t\,\cos t,$$

using the standard rules for derivatives of trigonometric functions.

1.3.1.2 *Binomial coefficients*

Here is a more difficult example: *Show that*

$$\sum_{j=0}^{n}\binom{n}{j} = 2^n,$$

where $\binom{n}{j}$ is the jth binomial coefficient: $\binom{n}{j} = \dfrac{n!}{j!\,(n-j)!}.$

Note that $n!$ or "n factorial" is $n(n-1)(n-2)\cdots 2\cdot 1$. We use the convention that $0! = 1$.

The formula to match is the binomial formula

$$(a + b)^m = \sum_{k=0}^{m} \binom{m}{k} a^k b^{m-k}.$$

Clearly we want to match m to n and k to j. But what about a and b? If we set $a = b = 1$ then $a^j = 1^j = 1$ and $b^{n-j} = 1^{n-j} = 1$ no matter what j and n are. Then we get

$$2^n = (1 + 1)^n = \sum_{j=0}^{n} \binom{n}{j}$$

as we wanted.

1.3.1.3   *Complex numbers and trigonometric functions*

A more sophisticated example using complex numbers is to match

$$1 + \cos \theta + \cos(2\theta) + \cdots + \cos(n\theta)$$

with the formula for a finite geometric series

$$1 + r + r^2 + \cdots + r^n = \frac{r^{n+1} - 1}{r - 1} \qquad \text{for } r \neq 1.$$

Trying to match r with $\cos \theta$ does not work since r^2 is not $\cos(2\theta)$. But there is a way to connect these: $e^{i\phi} = \cos \phi + i \sin \phi$ where $i = \sqrt{-1}$. Then $\cos \phi$ is the real part of $e^{i\phi}$; in fact $\cos(n\theta) = \operatorname{Re} e^{in\theta}$. Then we can write

$$1 + \cos \theta + \cos(2\theta) + \cdots + \cos(n\theta)$$
$$= \operatorname{Re} \left[1 + e^{i\theta} + e^{i2\theta} + \cdots + e^{in\theta} \right]$$
$$= \operatorname{Re} \left[1 + e^{i\theta} + (e^{i\theta})^2 + \cdots + (e^{i\theta})^n \right].$$

Aha! We can use $r = e^{i\theta}$. Then the finite geometric series formula becomes

$$1 + e^{i\theta} + (e^{i\theta})^2 + \cdots + (e^{i\theta})^n = \frac{(e^{i\theta})^{n+1} - 1}{e^{i\theta} - 1}$$
$$= \frac{e^{i(n+1)\theta} - 1}{e^{i\theta} - 1}.$$

Putting them together and using standard rules for complex numbers gives

$$1 + \cos \theta + \cos(2\theta) + \cdots + \cos(n\theta)$$
$$= \operatorname{Re} \frac{e^{i(n+1)\theta} - 1}{e^{i\theta} - 1} = \operatorname{Re} \frac{e^{i(n+1)\theta} - 1}{e^{i\theta} - 1} \frac{e^{-i\theta/2}}{e^{-i\theta/2}}$$
$$= \operatorname{Re} \frac{e^{i(n+1/2)\theta} - e^{-i\theta/2}}{e^{i\theta/2} - e^{-i\theta/2}}$$
$$= \operatorname{Re} \frac{\cos((n + \frac{1}{2})\theta) + i \sin((n + \frac{1}{2})\theta) - \cos(-\frac{1}{2}\theta) - i \sin(-\frac{1}{2}\theta)}{2 i \sin(\frac{1}{2}\theta)}$$
$$= \frac{\sin((n + \frac{1}{2})\theta) + \sin(\frac{1}{2}\theta)}{2 \sin(\frac{1}{2}\theta)} = \frac{1}{2} \left(\frac{\sin((n + \frac{1}{2})\theta)}{\sin(\frac{1}{2}\theta)} + 1 \right).$$

1.3.2 *Moving to the goal*

If we know where we want to get to, how do we get there? We need a way to reduce differences between where we are and what we want. The first step is to identify differences. Sometimes this involves complicating an expression rather than making it simpler, and this can seem unnatural. But if it gets us closer to what we want, all the better.

1.3.2.1 *Completing the square*

Can $a^2 + 2ab + 2b^2$ ever be negative? Not if we can write it as a sum of squares!

Now $(a + b)^2 = a^2 + 2ab + b^2$ (which matches the first two terms of $a^2 + 2ab + 2b^2$); subtracting this from $a^2 + 2ab + 2b^2$ we see that

$$a^2 + 2ab + 2b^2 - (a + b)^2 = a^2 + 2ab + 2b^2 - a^2 - 2ab - b^2 = b^2, \quad \text{so}$$
$$a^2 + 2ab + 2b^2 = (a + b)^2 + b^2.$$

Yes, $a^2 + 2ab + 2b^2$ is a sum of squares, so it can never be negative. There is even more here: the only way $a^2 + 2ab + 2b^2 = 0$ is if $a + b = 0$ and $b = 0$; that is, $a^2 + 2ab + 2b^2 = 0$ implies both $a = b = 0$.

1.3.2.2 *Calculus example: proving the product rule*

Now let's try to prove the product rule from calculus using the limit rules:

Theorem 1.3 (Product rule). *If u and v are differentiable at x, then*

$$\frac{d}{dx}(u(x)v(x)) = \frac{du}{dx}(x)\, v(x) + u(x)\, \frac{dv}{dx}(x).$$

Before we start the proof, we need to plan a strategy. We start from the left-hand side: $\frac{d}{dx}(u(x)v(x))$. We then unwrap the definition of derivative:

$$\frac{d}{dx}(u(x)v(x)) = \lim_{h \to 0} \frac{u(x + h)\, v(x + h) - u(x)\, v(x)}{h}.$$

Where do we want to get to? Let us unwrap the definition of the derivatives on the right-hand side:

$$\frac{du}{dx}(x)\, v(x) + u(x)\, \frac{dv}{dx}(x)$$
$$= \lim_{h \to 0} \frac{u(x + h) - u(x)}{h}\, v(x) + u(x)\, \lim_{h \to 0} \frac{v(x + h) - v(x)}{h}.$$

Then we want to connect these. Enough strategy. Now for the proof.

Proof. The left-hand side is

$$\frac{d}{dx}(u(x)v(x)) = \lim_{h \to 0} \frac{u(x+h)\,v(x+h) - u(x)\,v(x)}{h}.$$

We unwrapped the definition of derivative. If we look at our target, we want to get $u(x + h) - u(x)$ and $v(x + h) - v(x)$, so let's bring those things together: $u(x + h)\,v(x + h) = [u(x+h) - u(x)]\,v(x + h) + u(x)\,v(x + h)$. We will make this change, and hope that the other parts will work out.

Now,

$$u(x+h)\,v(x+h) - u(x)\,v(x)$$
$$= [u(x+h) - u(x)]\,v(x+h) + u(x)\,v(x+h) - u(x)\,v(x)$$
$$= [u(x+h) - u(x)]\,v(x+h) + u(x)\,[v(x+h) - v(x)].$$

This is a good sign: we have $v(x + h) - v(x)$ now.

So

$$\frac{d}{dx}(u(x)\,v(x))$$
$$= \lim_{h \to 0} \frac{u(x+h)\,v(x+h) - u(x)\,v(x)}{h}$$
$$= \lim_{h \to 0} \frac{[u(x+h) - u(x)]\,v(x+h) + u(x)\,[v(x+h) - v(x)]}{h}.$$

Now we want to get $(u(x + h) - u(x))/h$ and $(v(x + h) - v(x))/h$.

$$= \lim_{h \to 0} \frac{[u(x+h) - u(x)]\,v(x+h) + u(x)\,[v(x+h) - v(x)]}{h}$$
$$= \lim_{h \to 0} \left[\frac{u(x+h) - u(x)}{h}\,v(x+h) + u(x)\,\frac{v(x+h) - v(x)}{h} \right]$$
$$= \lim_{h \to 0} \frac{u(x+h) - u(x)}{h}\,\lim_{h \to 0} v(x+h)$$
$$\quad + u(x)\,\lim_{h \to 0} \frac{v(x+h) - v(x)}{h}$$
$$= \frac{du}{dx}(x)\,v(x) + u(x)\,\frac{dv}{dx}(x),$$

using the rules for limits, and continuity of $v(\cdot)$ (since it is differentiable at x). $\square$

1.3.3 *Dummy variables*

An issue that is often confusing to beginners is the role of dummy variables. Dummy variables are used as part of some formula or statement so that their value does not depend on any outside information. Dummy variables can occur in definitions, statements, sums, and integrals. Here is the definition of a function where x is a dummy variable:
$$f(x) = 1 + x^2 \qquad \text{for all } x.$$
If we had defined this using a different variable, but with the corresponding expression defining the function, we would be defining the same function:
$$g(y) = 1 + y^2 \qquad \text{for all } y.$$
Now f and g are exactly the same function: $f(0) = 1 + 0^2 = g(0)$, $f(1) = 1 + 1^2 = g(1)$, $f(37) = 1 + 37^2 = g(37)$, etc. In fact, $f(z) = 1 + z^2 = g(z)$ for all values of z so $f = g$ as functions.

1.3.3.1 *Dummy variables in statements*

Dummy variables also arise in statements with phrases like "*for all*" or "*there exists*". These are quantified statements, and the dummy variable is the variable to which the phrase applies. Quantified statements will be discussed in much more detail in Chapter 2 on logic. For example, consider the true statement
$$e^s \geq 1 + s \qquad \text{for all real } s.$$
Since this is true for all real s, we could replace s by any expression whose value is real, and the statement would still be true. In particular, we can replace s by any other variable and the statement would remain true.

On the other hand, consider the following statement, which also happens to be true:
$$\text{there exists } s > 0 \text{ where } e^s < 1 + 2s.$$
Numerical calculations show that this is true for $s = 0.4$. If we change the variable in this statement to t:
$$\text{there exists } t > 0 \text{ where } e^t < 1 + 2t$$
we still have a true statement (for example, with $t = 0.4$). We cannot replace s by a general expression and still expect to have a true statement, since it is a statement that there exists a value for which $e^s > 1 + 2s$. If we replace s by $2 + t^2$ then we can never get a real t where $2 + t^2 = 0.4$, so we cannot guarantee that the statement is true with this replacement. In fact the statement
$$\text{there exists } 2 + t^2 > 0 \text{ where } e^{2+t^2} < 1 + (2 + t^2)$$
is *false*. We can only guarantee the "there exists" statement is true if we replace the dummy variable with another dummy variable.

1.3.3.2 *Dummy variables in sums and integrals*

Dummy variables also occur in sums and integrals. Consider the expression

$$\sum_{k=1}^{10} k^2 = 1^2 + 2^2 + 3^2 + \cdots + 10^2.$$

If we replace k by some other (integer-valued) variable, we still get the same value of the expression:

$$\sum_{k=1}^{n} k^2 = \sum_{i=1}^{n} i^2 = \sum_{p=1}^{n} p^2 = 1^2 + 2^2 + \cdots + n^2.$$

In integrals the same kind of property holds:

$$\int_a^b f(x)\,dx = \int_a^b f(y)\,dy.$$

Dummy variables in sums and integrals are particularly useful in double sums and double integrals. First note that double sums with fixed lower and upper limits can have the sums exchanged:

$$\begin{aligned}
\sum_{i=1}^{m}\sum_{j=1}^{n} a_{ij} &= \sum_{i=1}^{m} \left[a_{i1} + a_{i2} + a_{i3} + \cdots + a_{in}\right] \\
&= a_{11} + a_{12} + a_{13} + \cdots + a_{1n} \\
&\quad + a_{21} + a_{22} + a_{23} + \cdots + a_{2n} \\
&\quad + a_{31} + a_{32} + a_{33} + \cdots + a_{3n} \\
&\quad\quad \vdots \\
&\quad + a_{m1} + a_{m2} + a_{m3} + \cdots + a_{mn} \\
&= a_{11} + a_{21} + a_{31} + \cdots + a_{m1} \quad\quad \text{(sum by columns)} \\
&\quad + a_{12} + a_{22} + a_{32} + \cdots + a_{m2} \\
&\quad + a_{13} + a_{23} + a_{33} + \cdots + a_{m3} \\
&\quad\quad \vdots \\
&\quad + a_{1n} + a_{2n} + a_{3n} + \cdots + a_{mn} \\
&= \sum_{j=1}^{n} \left[a_{1j} + a_{2j} + a_{3j} + \cdots + a_{mj}\right] \\
&= \sum_{j=1}^{n}\sum_{i=1}^{m} a_{ij}.
\end{aligned}$$

Then when we have symmetric double sums we can exchange the order of summation:

$$\sum_{i=1}^{n}\sum_{j=1}^{n} a_{ij} = \sum_{j=1}^{n}\sum_{i=1}^{n} a_{ij} \quad \text{(exchange sums)}$$

$$= \sum_{j=1}^{n}\sum_{k=1}^{n} a_{kj} \quad \text{(replace } i \text{ with } k)$$

$$= \sum_{i=1}^{n}\sum_{k=1}^{n} a_{ki} \quad \text{(replace } j \text{ with } i)$$

$$= \sum_{i=1}^{n}\sum_{j=1}^{n} a_{ji} \quad \text{(replace } k \text{ with } j).$$

At the end we have the same double sum, but we have swapped i and j in a_{ij}. This enables us to do some tricks with swapping dummy variables:

$$\sum_{i=1}^{n}\sum_{j=1}^{n} a_{ij} = \frac{1}{2}\sum_{i=1}^{n}\sum_{j=1}^{n} a_{ij} + \frac{1}{2}\sum_{i=1}^{n}\sum_{j=1}^{n} a_{ij}$$

$$= \frac{1}{2}\sum_{i=1}^{n}\sum_{j=1}^{n} a_{ij} + \frac{1}{2}\sum_{i=1}^{n}\sum_{j=1}^{n} a_{ji}$$

$$= \sum_{i=1}^{n}\sum_{j=1}^{n} \frac{1}{2}\left(a_{ij} + a_{ji}\right).$$

If $a_{ji} = -a_{ij}$, then $a_{ij} + a_{ji} = 0$ and so $\sum_{i=1}^{n}\sum_{j=1}^{n} a_{ij} = 0$.

A similar thing can be done with double integrals:

$$\int_{a}^{b}\int_{c}^{d} f(x,y)\,dy\,dx = \int_{c}^{d}\int_{a}^{b} f(x,y)\,dx\,dy = \int_{c}^{d}\int_{a}^{b} f(y,x)\,dy\,dx;$$

for the last equality we swap the roles of x and y just as we swapped the roles of i and j in the double sums above.

1.4   Proof by contradiction

One of the harder tricks to master in proofs is "proof by contradiction". In Calculus and your early mathematics courses, you mostly had to learn to calculate. You started from the expression you wanted to evaluate, and you proceeded step-by-step, simplifying, applying rules, until you couldn't simplify any more.

With proof by contradiction you need to work backwards. Suppose we want to prove "if p then q". Well, q can be true, or it can be false. It has to be one of the two possibilities. What is the worst that could happen? Well, q could be false. If q were false, we want to show that p must then also be false. The is the *contrapositive* statement "if not q then not p". This whole process can be repeated: "if not q then not p" implies that "if not (not p) then not (not q)", which is equivalent to "if p then q". The original statement and its contrapositive are equivalent.

So the strategy is this: assume that the conclusion (q) is false, and prove from this that the assumptions or hypothesis (p) must also be false. To make this clear, the following format will be used:

Theorem. *p implies q.*

Proof. **Suppose that the conclusion is false:** not q.

> *Body of the proof goes here.*

Therefore the hypothesis is false: not p.
   Thus: p implies q.   □

Here is a simple example:

Theorem 1.4. *If $x \neq 0$ then $x^2 \neq 0$.*

Proof. **Suppose that the conclusion is false:** Suppose $x^2 \neq 0$ is false.
   Then $x^2 = 0$, so $x = 0$.
   Therefore the hypothesis is false: since $x = 0$.
   Thus: $x \neq 0$ implies $x^2 \neq 0$.   □

Here is a more interesting example of proof by contradiction. It is a version of the "pigeon-hole" principle.

Theorem 1.5 (Generalized pigeon-hole principle). *If each of N items is put into exactly one of n pigeon-holes and $N > mn$, there is one pigeon-hole with at least $m + 1$ items.*

Proof. **Suppose that the conclusion is false:** That is, we suppose no pigeon-hole contains at least $m + 1$ items.
   Then each pigeon-hole contains $\leq m$ items. Since there are only n pigeon-holes, there can be at most only mn items. That is, $N \leq mn$.
   Therefore the hypothesis is false: since we assumed that $N > mn$.
   Thus: if $N > mn$ then at least one pigeon-hole has at least $m + 1$ items. □

The pigeon-hole principle is so simple and obvious it is easy to overlook. But it can be used in many places as a powerful tool (see Exercise 1.8).

Here is a more complex example of *proof by contradiction* for graphs. A basic introduction to graph theory can be found in Bollobás (1979).

Definition 1.3. *A graph $G = (V, E)$ is a collection of* vertices *or* nodes *V, together with a set of* edges *E that connect the nodes of the graph. In* undirected *graphs, each edge is a set of exactly two vertices which are the endpoints of the edge. In* directed *graphs, each edge has two endpoints: a start node and an finish node. A* loop *is an edge where the start and finish nodes are the same. The* degree *of a node is the number of edges of which it is an endpoint. If a node n is the endpoint of an edge e we say that n is* incident *to e.*

In a directed graph, a pair of nodes can have more than one edge between them and an edge can start and end at the same node.

Definition 1.4. *A* path *in a graph $G = (V, E)$ is a sequence*

$$n_1, e_1, n_2, e_2, \ldots, e_k, n_{k+1}$$

where each $n_j \in V$ is a node or vertex of G, each $e_j \in E$ is an edge of G, and n_j and n_{j+1} are the start and finish nodes of e_j respectively. In an undirected graph, we just require that $\{n_j, n_{j+1}\}$ is the set of endpoints of e_j so $n_j \neq n_{j+1}$. A simple path *is a path where no edge or node is repeated, except possibly $n_1 = n_{k+1}$, in which case the path is called a* cycle. *A graph is* connected *if for every two nodes x and y there is a path joining them $x, e_1, n_2, \ldots, e_k, y$. If a graph is not connected, we say that it is* disconnected.

Some examples of graphs are shown in Figure 1.1. The undirected graph is connected, while the directed graph is not (no path can end at the node with the loop).

Theorem 1.6. *If a finite connected undirected graph G has no cycles then G has a node with degree equal to one.*

The contrapositive to "G has no cycles implies G has a node with degree one" is "G has no node with degree one implies G has a cycle". To help with the proof we can assume that G is finite and connected. Connectedness implies that no node has degree zero (which would immediately mean the node is isolated). So we can replace the contrapositive with "G has no node with degree ≤ 1 implies that G has a cycle". So the aim is to construct a cycle given that every node has degree ≥ 2.

Proof. We assume that G is finite and connected.

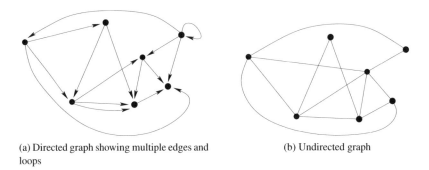

(a) Directed graph showing multiple edges and loops

(b) Undirected graph

Fig. 1.1: Directed and undirected graphs

> *The proof is by contradiction, showing that if the conclusion "G has a node with degree one" is false, then the hypothesis "G has no cycle" is false.*

Suppose that the conclusion is false: G has no node with degree equal to one.

Since G is connected, no node of G can have degree zero. Thus all nodes of G must have degree ≥ 2.

We now show that a cycle exists in G. To do that we construct a path and show that eventually there must be two identical nodes in the path.

> *The construction is a step-by-step or inductive construction. It stops as soon as a cycle is formed.*

Start with any node n_0 in G. Since $\deg(n_0) \geq 2$, there must be an edge e_0 incident to n_0. Let n_1 be the other end-point of e_0. Now $\deg(n_1) \geq 2$, so there must be another edge e_1 incident n_1. We continue with this construction: Suppose that so far we have constructed the path $n_0, e_0, n_1, \ldots, e_{j-1}, n_j$ where $n_k \neq n_\ell$ for $k \neq \ell$ and $0 \leq k < \ell \leq j$. Since $n_j \neq n_\ell$ for any $0 \leq \ell < j$, e_{j-1} is the only edge in the path incident to n_j. Since $\deg(n_j) \geq 2$, there must be at least one other edge e_j incident to n_j. Let n_{j+1} be the other node incident to e_j. If $n_{j+1} = n_\ell$ for some $0 \leq \ell \leq j$, then $n_\ell, e_\ell, n_{\ell+1}, \ldots, e_{j-1}, n_j, e_j, n_{j+1} = n_\ell$ is a cycle, as we wanted. Otherwise, we can continue with the construction of a path with distinct nodes.

We cannot continue adding to the path forever since G is finite: the number of different nodes in the path must be less or equal to the number of nodes in G. Thus for some j we must have a node $n_{j+1} = n_\ell$ for some $0 \leq \ell \leq j$, and we have constructed a cycle in G.

Therefore the hypothesis is false: we have proved that G must have a cycle.

Thus: if G is a finite connected graph with no cycles, G must have a node with degree one. □

Two comments should be made about this proof:

(1) The construction can be made more formal by using mathematical induction. See Section 2.6 for more information.
(2) We had to assume that G is finite in order to show that the path could not continue forever. Is there an infinite graph G which has no cycles but all nodes have degree ≥ 2? The answer is yes, but you will need to determine a suitable graph yourself. See Exercise 1.14.

1.5 "If and only if"

If p and q are two statements, then we say "p *if* q" for "q *implies* p" and "p *only if* q" for "p *implies* q". Combining these, "p *if and only if* q" means "p *implies* q, and q *implies* p". That is, "p *if and only if* q" means that p and q are logically equivalent: if one is true, so is the other; if one is false, so is the other. Sometimes, "p *if and only if* q" is abbreviated by "p *iff* q". Proving logical equivalence of two statements can be done by either a sequence of simpler logical equivalences (p *if and only if* q, and q *if and only if* r, shows that p *if and only if* r) or by a pair of implications (p *implies* q, and q *implies* p).

Note that the *converse* of "p *implies* q" is "q *implies* p". Thinking that a statement and its converse are logically equivalent is, of course, completely wrong and must be avoided: "*If it is raining, I take an umbrella*" might be true, but it does *not* mean that "*If I take an umbrella, then it is raining.*"

Here is an example of using a pair of implications to prove "*if and only if*".

Theorem 1.7. *If x is a real number, then $x \neq 0$ if and only if $x^2 > 0$.*

Proof.

> *First we show $x \neq 0$ implies $x^2 > 0$. We do this by looking at the two cases $x > 0$ and $x < 0$ separately.*

First suppose that $x \neq 0$. We show that then $x^2 > 0$.

If $x \neq 0$ then either $x > 0$ or $x < 0$. If $x > 0$ then $x^2 = x\,x$ is a product of positive numbers, and so is positive; that is $x^2 > 0$. If $x < 0$ then $x^2 = x\,x$ is a product of two negative numbers, which is also positive; that is, $x^2 > 0$. In either case, $x^2 > 0$.

So $x \neq 0$ implies $x^2 > 0$.

Now for the converse...

We now show the converse: $x^2 > 0$ implies $x \neq 0$. This is done using proof by contradiction.

Suppose that the conclusion is false: that is, suppose that $x \neq 0$ is false. Then $x = 0$.

Then $x^2 = 0 \times 0 = 0$.

Therefore the hypothesis is false: x^2 is not positive, which is a contradiction.

Thus: $x^2 > 0$ implies $x \neq 0$.

Now we combine the two implications.

Since $x \neq 0$ implies $x^2 > 0$, and $x^2 > 0$ implies $x \neq 0$, the two statements $x \neq 0$ and $x^2 > 0$ are logically equivalent.   □

Note that the arguments for the two implications ($x \neq 0$ implies $x^2 > 0$) and ($x^2 > 0$ implies $x \neq 0$) are quite different. Often there is an easy implication and a hard implication.

Showing that many statements are logically equivalent is most efficiently done by showing that there is a cycle of implications. For example, to show that statements p, q, r and s are logically equivalent, it is enough to show that p implies q, q implies r, r implies s, and s implies p. Proving that each pair is logically equivalent separately would involve proving twelve implications!

1.6   Drawing pictures

> *Geometry is the science of correct reasoning on incorrect diagrams.*
>
> George Pólya

Figures are a great aid to understanding. Sometimes it seems that figures by themselves are the proof! This is not quite true, and sometimes figures can be misleading, especially if not drawn correctly.

1.6.1   *Pythagoras' theorem — picture version*

The outer figure $ACEG$ in Figure 1.2 is a square of side $a + b$. There are some things to check about these diagrams. In particular, we need to check that the angles $\angle HBD$, $\angle BDF$, $\angle DFH$ and $\angle FHB$ are indeed right angles. The strategy is to compute the area of the outer square minus the area of the right angle triangles, which gives the area of $BDFH$, which we show to be a square.

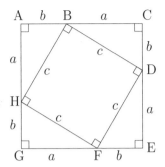

Fig. 1.2: Pythagoras' theorem

Theorem 1.8. *If a, b are the lengths of the edges adjacent to the right angle in a right angle triangle, and c is the length of the remaining edge (the hypotenuse), then*

$$c^2 = a^2 + b^2.$$

Proof. In Figure 1.2, note that the triangles $\triangle ABH$, $\triangle CDB$, $\triangle EFD$, and $\triangle GHF$ are all congruent as they all have sides of length a and b and have a right angle between them (using the side-angle-side rule).

The area of the square $ACEG$ is $(a + b)^2 = a^2 + 2ab + b^2$. The area of each of the four congruent triangles is $\frac{1}{2}ab$, so the total area of the four triangles is $4 \times \frac{1}{2}ab = 2ab$. Thus the remaining area is $(a + b)^2 - 2ab = a^2 + b^2$. We now need to check that $BDFH$ really is a square. All the lengths of its edges are the same (c) since the edges of $BDFH$ are all edges of congruent triangles. But we need to check that $\angle HBD$, $\angle BDF$, $\angle DFH$ and $\angle FHB$ are all right angles, as otherwise $BDFH$ could be a parallelogram.

We can see that $BDFH$ is a square in two ways: Rotating the square $ACEG$ by 90° leaves the diagram unchanged, so all of $\angle HBD$, $\angle BDF$, $\angle DFH$ and $\angle FHB$ must be the same. The sum of the interior angles of a quadrilateral must be 360° so each of them must be 90°. An alternative approach is to note that $\angle BDC$ and $\angle EDF$ are complementary angles of congruent right angle triangles, and so sum to 90°. Since $\angle CDE$ is 180°, it follows that $\angle BDF$ is 90°. This argument applies to all of the angles $\angle HBD$, $\angle BDF$, $\angle DFH$ and $\angle FHB$, so all are 90°.

Thus $BDFH$ is a square, and so has area c^2. Equating the area computed gives $c^2 = a^2 + b^2$. □

Note that this is an example of

computing the same thing in two different ways.

An alternative proof of Pythagoras' theorem is to re-arrange the triangles in Figure 1.2 to form Figure 1.3. The area of the shaded part of Figure 1.3 is clearly $a^2 + b^2$, but it should also be equal to the area of the shaded part of Figure 1.2, which is c^2.

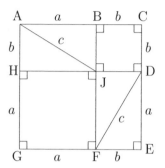

Fig. 1.3: Alternative proof of Pythagoras' theorem

1.6.2 *Eulerian paths and the bridges of Koenigsberg*

Leonhard Euler (1707–1783) heard about the people of Koenigsberg (now a city called Kaliningrad) who played a game of trying to cross each and every bridge in the city exactly once. No-one could do it, and Euler set out to find out why.

Figure 1.4 is a map of the old city of Koenigsberg, with the bridges and the river Pregel highlighted. As far as bridge crossings are concerned, the connected parts of dry land can be collapsed down to single points: a person can walk between any two points on the same connected piece of dry land without crossing a bridge. These points are connected by the bridges, which become edges connecting the points. This leaves us with a much more idealized picture of the problem (Figure 1.5).

Figure 1.5 simplifies the problem. In particular, it *abstracts* the essential items from the mass of distracting detail shown in Figure 1.4. We have not yet begun to write a proof, but this step is very important:

simplify a task down to its essential elements

The task the Koenigsbergers set themselves was to cross each bridge exactly once in a single path. In Figure 1.5, this corresponds to finding a path through the network or graph which crosses each edge exactly once. Definitions for graph,

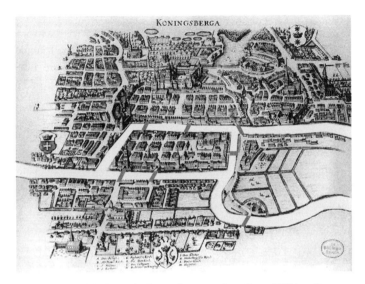

Fig. 1.4: Old Koenigsberg. Image taken from Wikipedia.

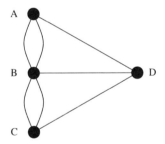

Fig. 1.5: Idealized old Koenigsberg.

path, degree, and so on can be found in Definitions 1.3 and 1.4 on page 15. We say a path (or cycle) is and Euler path (or cycle) if it contains each edge of the graph exactly once. Note that an Euler cycle may contain any given node multiple times, but each edge will occur exactly once.

The theorem we will prove is the following:

Theorem 1.9. *Suppose we have a connected undirected graph. If the degree of every node is even, there is an Euler cycle; if the degree of all but two nodes is even, there is an Euler path and starts and ends at the two nodes of odd degree. If the number of nodes with odd degree is neither zero nor two, then there is no Euler path (or cycle).*

The proof of this theorem will be delayed until Section 3.7; there are some more techniques that we need to complete the proof.

1.7   Notation

Notation helps us express our thoughts, and so is important. There are certain conventions that mathematicians tend to use, and we generally abide by. (For example, if we refer to a number π, it had better be $3.1415926\ldots$, or it will be very confusing.) But there are few conventions. Generally, a mathematician can choose notations mostly as she or he pleases. But you must be sure to define any notation that is not already very well known.

1.7.1   *Choosing names*

In choosing names of variables, or functions, we can follow traditions. For example, inputs to functions tend to be x or t, but can be r or s or α. Functions tend to be f or g. We can choose from the Roman alphabet a, b, c, d, $\ldots$, A, B, C, D, $\ldots$ or the Greek alphabet α, β, γ, δ, $\ldots$, A, B, Γ, Δ, $\ldots$.

Try to avoid using symbols together that are too similar, such as i (Roman "i") and ι (Greek "iota"), or Δ (Greek "Delta") and $\triangle$ (a triangle). Some symbols can be confusing in certain contexts, such as the Greek capital Xi (Ξ). The ratio of Ξ to its complex conjugate $\overline{\Xi}$ is an expression

$$\frac{\Xi}{\overline{\Xi}}$$

that can be hard to decipher, especially when written on a blackboard! Sometimes other languages or scripts are raided for new symbols: ⋔ (pitchfork), ℶ ("beth" from Hebrew), or ð ("thorn" from Old English or Icelandic). But these should only be used as a last resort, and for concepts that have a permanent meaning, such as $\aleph_0$ (the first infinite cardinal).

One rule is paramount:

   do not use the same name for different things!

You *can* use *different names for the same thing*, and in fact you may need to before you know that they are the same thing. For example, in proving uniqueness of solutions of a differential equation

$$\frac{dy}{dt} = f(t, y), \qquad y(t_0) = y_0,$$

we need to start with a statement like "*Suppose y_1 and y_2 are solutions of the differential equation...*" and end with "*... so $y_1(t) = y_2(t)$ for all t*". Yes, y_1 and y_2 are the same, but we did not know that at the beginning.

Many mathematicians have a system for choosing names: Greek lower case letters α, β, γ, ... for scalars, Roman bold lower case letters for vectors **x**, **y**, **z**, ..., and Roman upper case letters A, B, C, ... for matrices. Sometimes for closely related variables, one might be a Roman letter and the other the corresponding Greek letter, such as x and ξ (xi), or t and τ (tau).

1.7.2 *Introducing new notation*

Introducing new notation is like introducing a new character in a story. We need to do it, but doing too much is distracting and confusing. Remember, every new notation is a burden on the reader. But we do this to lighten the reader's other burdens, and prevent confusion.

For example, replacing a variable by an expression *without* that variable is usually less confusing than replacing a variable with an expression that contains that variable. Consider replacing "x" in "$f(x) = 3x^2 + 2x + 1$" by "$7x - 2$": this gives "$f(7x - 2) = 3(7x - 2)^2 + 2(7x - 2) + 1$". Now try replacing "$z$" in "$f(z) = 3z^2 + 2z + 1$" by "$7x - 2$". Again we get "$f(7x - 2) = 3(7x - 2)^2 + 2(7x - 2) + 1$". But it is usually easier to follow what happens if the variable is not in the expression.

Here is another example.

Definition 1.5. *A square matrix A is* positive semi-definite *if for every compatible vector* **x**, $\mathbf{x}^T A \mathbf{x} \geq 0$.

Theorem 1.10. *If A is a positive semi-definite $n \times n$ matrix, and B is any $n \times p$ matrix, then $B^T AB$ is a positive semi-definite matrix.*

Proof. We want to show that $\mathbf{z}^T B^T AB\mathbf{z} \geq 0$ for any **z**.

> *This unwraps the definition of positive semi-definite.*
> *But why use* **z**, *and not* **x**?

If $\mathbf{x} = B\mathbf{z}$, then $\mathbf{z}^T B^T AB\mathbf{z} = (B\mathbf{z})^T A(B\mathbf{z}) = \mathbf{x}^T A\mathbf{x} \geq 0$, since A is positive semi-definite.

> *Using* **z** *instead of* **x** *enables us to set* $\mathbf{x} = B\mathbf{z}$.

Since $\mathbf{z}^T B^T AB\mathbf{z} \geq 0$ for all **z**, $B^T AB$ is positive semi-definite.

$\square$

We cannot write "*Let* $\mathbf{x} = B\mathbf{x}$." That would be nonsense. We could write "*Replace* $\mathbf{x}$ *by* $B\mathbf{x}$ *in* $\mathbf{x}^T A\mathbf{x}$." But this requires the reader to do some mental gymnastics. The better solution is to introduce a new variable $\mathbf{z}$. It is especially easy to do since the "$\mathbf{x}$" in Definition 1.5 is a dummy variable.

1.7.3 *Sub-scripts, super-scripts, and decorations*

Often we first see sub-scripts for entries in a vector: x_3 is the 3rd entry of a vector $\mathbf{x}$. Super-scripts are first used for powers: $x^3 = x\,x\,x$. But we can quickly have trouble keeping these straight. For example, we might not want subscripts on a vector $\mathbf{x}_3$ because we would first think that it is the 3rd entry of $\mathbf{x}$. So we might use super-scripts for a sequence of vectors: $\mathbf{x}^1$, $\mathbf{x}^2$, $\mathbf{x}^3$, ... Then the kth entry of the nth vector in the sequence is x_k^n. But is this the nth power of the kth entry of a single vector $\mathbf{x}$?

Be careful about over-using sub-scripts and super-scripts. It might, for example, be better to write $(\mathbf{x}^n)_k$ to mean the kth entry of the nth vector in the sequence.

Other decorations are used, sometimes to denote an operation performed on the decorated object, sometimes to indicate that the variable is a certain kind of object, and sometimes simply as a way of creating a new (but usually related) variable. For example, it is common for $\widehat{f}$ to denote the Fourier transform a function f, but $\widehat{x}$ might be the computed estimate of a quantity whose actual value is x, or x_3 is the 3rd entry of a vector $\vec{x}$. Usually, if a decoration is used simply to indicate that the variable is of a particular kind, then the decoration should be only used for that purpose.

You should aim for consistency in how you use decorations. For example, if $\widehat{x}$ is a computed estimate of x, then $\widehat{y}$ should be a computed estimate for y.

1.8 More examples of proofs*

1.8.1 *Composition of continuous functions*

This example will give us more practice at unwrapping definitions and dealing with dummy variables.

Theorem 1.11. *If f and g are continuous real-valued functions of a real variable, so is the composition $h(x) = f(g(x))$.*

Before we start, we should recall the ϵ-δ definition of continuity: a function q

is continuous means that for every z_0 and $\epsilon > 0$ there is a $\delta > 0$ where

$$|z - z_0| < \delta \text{ implies } |q(z) - q(z_0)| < \epsilon.$$

Remember that ϵ, δ, z and z_0 are dummy variables: we may need to use different names for the different applications of the definition (f is continuous, g is continuous, and for showing that h is continuous). To show continuity of h we want to show that for every x_0 and $\epsilon_h > 0$ there is a $\delta_h > 0$ where

$$|x - x_0| < \delta_h \text{ implies } |h(x) - h(x_0)| < \epsilon_h.$$

On the other hand, we know that f and g are continuous: for every y_0 and $\epsilon_f > 0$ there is a $\delta_f > 0$ where

$$|y - y_0| < \delta_f \text{ implies } |f(y) - f(y_0)| < \epsilon_f,$$

and for every x_0 and $\epsilon_g > 0$ there is a $\delta_g > 0$ where

$$|x - x_0| < \delta_g \text{ implies } |g(x) - g(x_0)| < \epsilon_g.$$

This might seem like a lot of new variables, but giving them different names means we have more freedom in how to choose them. (An exception: The x and x_0 we use for g are the same as the x and x_0 we use for h, since these must be the same.) By delaying making decisions until we have more information we can make better choices.

Proof. Since f and g are continuous, for every x_0 and $\epsilon_g > 0$ and y_0 and $\epsilon_f > 0$, there are $\delta_g > 0$ and $\delta_f > 0$ where

$$|x - x_0| < \delta_g \text{ implies } |g(x) - g(x_0)| < \epsilon_g,$$
$$|y - y_0| < \delta_f \text{ implies } |f(y) - f(y_0)| < \epsilon_f.$$

Unwrapping the definitions is a good starting point. Now we need to look to our destination, by unwrapping the definition of h.

Suppose $\epsilon_h > 0$ is given. Then we want to show that there is a $\delta_h > 0$ where

$$|x - x_0| < \delta_h \text{ implies } |h(x) - h(x_0)| = |f(g(x)) - f(g(x_0))| < \epsilon_h.$$

Now we need to try matching up the different ϵ*'s and* δ*'s.*

Let $\epsilon_f = \epsilon_h > 0$. Then we have $\delta_f > 0$ where $|y - y_0| < \delta_f$ implies $|f(y) - f(y_0)| < \epsilon_f = \epsilon_h$.

Matching variables between what we have and what we want means we set $y = g(x)$ *and* $y_0 = g(x_0)$. *This shows us that we should have* $\epsilon_g = \delta_f$.

Let $\epsilon_g = \delta_f > 0$. Then there is a $\delta_g > 0$ where $|x - x_0| < \delta_g$ implies $|g(x) - g(x_0)| < \epsilon_g = \delta_f$. Then setting $y = g(x)$ and $y_0 = g(x_0)$ we have $|x - x_0| < \delta_g$ implies $|g(x) - g(x_0)| = |y - y_0| < \epsilon_g = \delta_f$, and so $|f(g(x)) - f(g(x_0))| = |f(y) - f(y_0)| < \epsilon_f = \epsilon_h$.

Then putting $\delta_h = \delta_g$ we see that $|x - x_0| < \delta_h$ implies $|h(x) - h(x_0)| < \epsilon_h$, as we wanted. $\square$

Unwrapping the definitions early on gets us started, but it is easier if we give all the dummy variables different names; after all, the δ or the ϵ we use for f may be different from the δ or ϵ we use for g. Naming them the same would force them to have the same value, which we do not want.

1.8.2   *Sums of squares*

Here is a theorem from number theory:

Theorem 1.12. *No number of the form $4m + 3$, with m a whole number, can be written as the sum of exactly two squares.*

You can try for yourself: start with $m = 5$ so $4m + 3 = 23$ and try writing 23 as a sum of two squares.

Since we are asked to prove a negative, proof by contradiction is a natural method to do so.

Proof. **Suppose that the conclusion is false:** Suppose that $4m + 3 = x^2 + y^2$ with x and y whole numbers.

Now we have to look at various cases; in particular, whether x is even or odd, and whether y is even or odd.

Case I: Suppose both x and y are even. Then $x = 2k$ and $y = 2\ell$ for some whole numbers k and ℓ. Then $x^2 + y^2 = 4k^2 + 4\ell^2 = 4(k^2 + \ell^2)$ which has remainder zero when divided by 4, and so cannot be $4m + 3$.

Case II: Suppose x is even but y is odd. Then $x = 2k$ and $y = 2\ell + 1$ for some whole numbers k and ℓ. Then $x^2 + y^2 = 4k^2 + (2\ell + 1)^2 = 4k^2 + 4\ell^2 + 4\ell + 1 = 4(k^2 + \ell^2 + \ell) + 1$ which has remainder one when divided by 4, and so cannot be $4m + 3$.

Case III: Suppose x is odd but y is even. Swapping x and y, we get **Case II**, which is impossible.

Case IV: Suppose that x and y are odd. Then $x = 2k + 1$ and $y = 2\ell + 1$ for some whole numbers k and ℓ. Then $x^2 + y^2 = (2k + 1)^2 + (2\ell + 1)^2 =$

$4k^2 + 4k + 1 + 4\ell^2 + 4\ell + 1 = 4(k^2 + k + \ell^2 + \ell) + 2$ which has remainder two when divided by 4, and so cannot be $4m + 3$.

No matter which case we consider, it is impossible to have $x^2 + y^2 = 4m + 3$.

Therefore, $4m + 3$ cannot be written as the sum of two squares. $\square$

1.9 Exercises

(1) *Changing what you have to look like what you want.* Prove the quotient rule from calculus using the rules for limits. That is, show that

$$\frac{d}{dx}\left(\frac{f(x)}{g(x)}\right) = \frac{f'(x)\,g(x) - f(x)\,g'(x)}{g(x)^2},$$

provided f and g are differentiable at x and $g(x) \neq 0$. [**Hint:** Write $\dfrac{f(x+h)}{g(x+h)} - \dfrac{f(x)}{g(x)}$ with a common denominator.]

(2) In many cases, working out where you are trying to get to is the hardest part. For example, how could we figure out the quotient rule for the previous question? One way is to write $q(x) = f(x)/g(x)$ and *suppose* that $q(x)$ is differentiable. We want to find a formula for $q'(x)$ using only the product rule. Write $q(x)\,g(x) = f(x)$ and differentiate both sides using the product rule; solve what comes out for $q(x)$ to get the quotient rule above.

(3) *Handling cases.* A *non-decreasing* function is a function f where $x \geq y$ implies $f(x) \geq f(y)$. Prove rigorously and without calculus that $f(z) = \max(0, z)^2$ is an non-decreasing function. [**Hint:** There are three cases: $x, y \geq 0$, $x \geq 0 > y$, and $0 > x, y$. The case $y \geq 0 > x$ does not occur if $x \geq y$.]

(4) Prove that for any real number x, $\sqrt{1 + x^2} > x$.

(5) Prove that $\sqrt{3}$ is an irrational number. [**Hint:** You can use the proof of the irrationality of $\sqrt{2}$ as a model. You will come to the result that $3b^2 = a^2$, from which we want to show that a is divisible by three. You may need to consider three cases: $a = 3c$, $a = 3c + 1$ or $a = 3c + 2$ for some integer c.]

(6) *Symmetry.* Show that $\sum_{k=1}^{n-1}\cos(k\pi/n) = 0$ where n is a positive integer. [**Hint:** Put $\ell = n - k$ to re-write the sum, and use the fact that $\cos(\pi - \theta) = -\cos\theta$ for all θ. You should be able to show that the sum is equal to its negative.]

(7) *Proof by contradiction.* Prove that $e = \sum_{n=0}^{\infty} 1/n!$ is irrational. Do so by first supposing that e is rational so we can write $e = a/b$ with a and b positive integers, $b \neq 0$. Then $b!\,e = ((b-1)!)\,(b\,e) = (b-1)!\,a$ would be an integer. Show that $b!\,e$ is *not* an integer by looking at $b!\sum_{n=b+1}^{\infty} 1/n!$.

(8) *Proof by contradiction.* Suppose we have a collection $\{x_1, x_2, \ldots, x_{21}\}$ of 21 different integers in the range zero through 100. Show that there are at least two pairs of integers with the same sum.

(9) *Changing what you have to look like what you want.* Prove that if f is differentiable at a, then it is continuous at a. [**Hint:** We are given that $\lim_{x \to a}(f(x) - f(a))/(x - a) = f'(a)$ and we want to show that $\lim_{x \to a} f(x) = f(a)$. So we try to write $f(x)$ in a way that involves $(f(x) - f(a))/(x - a)$. Then taking limits gives the desired result.]

(10) *Changing what you have to look like what you want.* Let $f(x) = \int_1^x z^{-1}\, dz$ for $x > 0$. (This is, in fact, the natural logarithm function.) Show that for x, $y > 0$ we have $f(xy) = f(x) + f(y)$. [**Hint:** Write $f(xy) = f(x) +$ something and unwrap the definitions; then show that the "something" is $f(y)$.]

(11) *Unwrapping definitions.* The squeeze principle says that if $a_n \le c_n \le b_n$ for all n and $\lim_{n \to \infty} a_n = \lim_{n \to \infty} b_n = L$, then $\lim_{n \to \infty} c_n = L$. The first step of proving this is to unwrap the definition of limit for $\lim_{n \to \infty} a_n = L$, $\lim_{n \to \infty} b_n = L$, and the objective $\lim_{n \to \infty} c_n = L$. Prove the squeeze principle by first unwrapping these definitions. Given $\epsilon > 0$ find a suitable choice for N so that $n \ge N$ implies $|c_n - L| < \epsilon$.

(12) *Drawing pictures.* Draw a plot of $f(x) = e^x$. We can think of this as an example of a convex function ($0 \le \theta \le 1$ implies that $f(\theta x + (1 - \theta)y) \le \theta f(x) + (1 - \theta)f(y)$). Now consider what happens as y approaches x from above: the slopes of the chord lines decrease. Prove that this is true for any convex function.

(13) *"If and only if."* Show that $0 \le x \le 1$ if and only if $x^2 \le x$. Clearly show the forward implication ($0 \le x \le 1$ implies $x^2 \le x$), and the converse ($x^2 \le x$ implies $0 \le x \le 1$).

(14) ⚠ In connection with Theorem 1.6, try to create an *infinite* graph G that has no cycles, but also no nodes of degree one or zero.

(15) *Delay making decisions.* Prove Theorem 1.2. As in Theorem 1.11, use different symbols for each function until you have enough information to choose them.

(16) Students in calculus often give solutions to homework without words. What would a proof look like if it was done that way? Here is a possibility: (Turn to next page.)

Theorem. $\sqrt{2}$ *is irrational.*

"Proof".

$$\sqrt{2} = a/b \qquad \text{no c.f.}$$
$$2b^2 = a^2$$
$$a \text{ even} \ : \ 2b^2 = (2c)^2 = 4c^2$$
$$b^2 = 2c^2 \Rightarrow b \text{ even} \qquad \times$$
$$a \text{ odd} \ : \ 2b^2 = (2c+1)^2 = 4(c^2+c) + 1 \qquad \times$$
$$\text{Contradiction!}$$

"□"

Which is more readable? Try doing this with one or more proofs from this chapter. Can you understand the result? Can your friends/classmates understand the result? How important are words and sentences in proofs?

Chapter 2

Logic and other formalities

Proofs are meant to be logical, but what exactly is logic? And do we use it in proofs?

Logic is meant to be part of the foundation of mathematics. And mathematics can be translated into pure logic.

The ancient Greeks had a kind of formal logic known as *syllogisms*. These were rules for inferring true statements from a pair of different true statements. For example,

> *All men are mortal.*
> *Aristotle is a man.*
> Therefore: *Aristotle is mortal.*

However, syllogisms are limited in their complexity even though there were 11 different types. Mathematics needs a more sophisticated kind of formal logic. This was developed first only in the early-1900's based on the work of George Boole and August De Morgan, and led to the work of Gottlob Frege, Giuseppe Peano, culminating in the book *Principia Mathematica* Whitehead and Russell (1925–1927).

Essentially all of mathematics can be translated into the system of Russell and Whitehead's *Principia Mathematica*. But even the simple statement that "$1 + 1 = 2$" does not appear until after page 360. While everything is made rigorous in *Principia Mathematica*, it is not very practical.

Why then should we use formal logic? Because it ensures that our reasoning is properly founded. Without it, there are many tempting *but false* inferences we might make. By going back to logic we can check our reasoning to make sure that our inferences are valid. In this chapter we give a brief overview of the basics of mathematical logic, starting with propositional calculus and leading to predicate calculus.

We begin with *propositional calculus*. This is the logic of statements, without going into why a particular statement is true or false. These statements can be combined by using logical operations "and", "or", "not", "implies", and "equivalence". Basic statements combined in this way form an algebra called a *Boolean algebra*.

The next step is to add variables, expressions and predicates. Predicates are functions with values that are either true or false: for example a predicate $P(x)$ could be "$x > 3$". If $x = 4$ then $P(x)$ is true; if $x = 2$ then $P(x)$ is false. Along with variables, expressions and predicates we need quantifiers "$\forall$" and "$\exists$". If $s(x)$ is a statement in which x is a variable, then we have the statements "$\forall x\, s(x)$", which means "*for all x the statement $s(x)$ is true*", and the statement "$\exists x\, s(x)$" which means "*there is a value of x for which the statement $s(x)$ is true*". Statements can also be combined just as in propositional calculus using "and" ($\wedge$), "or" ($\vee$), "not" ($\neg$), "implies" ($\Rightarrow$), and "equivalence" ($\Leftrightarrow$). The resulting system is called *predicate calculus*.

Here we will review propositional and predicate calculus, and how we use them in mathematical proofs. The proofs that we usually write down are not proofs in formal logic; the logic we use in practice is informal logic. But we aim to make our proofs rigorous enough that they *could* be translated into formal logic as correct proofs.

In this chapter we will also look at the concepts of "sets," "functions," and "algorithms." Sets and functions are typically covered in most textbooks on logic or discrete structures. Proving algorithms correct via logic is discussed in some discrete structures textbooks, such as (Gersting, 2007, §§1.6, 2.3).

2.1 Propositional calculus

2.1.1 *Propositional logic and truth tables*

In propositional logic basic statements (or propositions) are represented by simple variables p, q, r, etc. For example, a variable p could represent the statement "*It rained here today*," and q could represent "*Three is bigger than two*." Propositions can be combined and modified by the following logical connectives: "not" ($\neg$ or $\sim$), "and" ($\wedge$ or $\&$), "or" ($\vee$), "implies" ($\to$ or $\Rightarrow$), and "equivalent to" ($\leftrightarrow$ or $\Leftrightarrow$ or $\equiv$). Expressions can and should be parenthesized.

Here is how they are interpreted:

- "$\neg p$" means "*It did not rain here today.*"

Double negation ("not not") can be a bit difficult to follow sometimes; in mathematics double negation is a positive. Sometimes in everyday English, a double negative is emphatically negative, but that is not the mathematical meaning.

- "$p \wedge q$" means "*It rained here today and three is bigger than two.*"
- "$p \vee q$" means "*Either it rained today or three is bigger than two, or both are true.*"

  The mathematician's version is called *inclusive or*. The other version "*Either it rained today or three is bigger than two, but **not** both*" is called *exclusive or*. Sometimes in English, exclusive or is implied, often by tone of voice: mother to 5-year old "*You can have the ice cream **or** the soda...*" Note that exclusive or is often denoted by "$\underline{\vee}$" or "$\oplus$", and $p \underline{\vee} q$ is equivalent to $\neg(p \Leftrightarrow q)$.

- "$p \Rightarrow q$" means "*If it rained here today, then three is bigger than two.*"

  This can be expressed as "*if p then q*", "*p implies q*", "*q if p*", or "*p only if q*".

  The biggest trap that beginners fall into is to think that "$p \Rightarrow q$" means that p has to somehow *cause* q. Telling if one thing causes some other thing can be an unbelievably difficult task. For "*If Chinese new year falls on a Sunday, the stock market will rise*", we might be able to tell if the stock market does indeed rise every year the Chinese new year falls on a Sunday. But how could you tell if the day of the Chinese new year *caused* the stock market to rise? How could you tell that the day of the Chinese new year *did not cause* the stock market to rise? Who knows... there may be mysterious forces linking these two events that we do not understand.

  Mathematicians, being much more practical people, just want to tell if "$p \Rightarrow q$" is true just be looking up the truth of p, the truth of q, and checking a truth table.

- "$p \Leftrightarrow q$" means "*It rained here today if and only if three is bigger than two.*"

  This can be expressed as "*p if and only if q*", or "*p and q are equivalent*".

  Usually, "$p \Leftrightarrow q$" is defined as "$(p \Rightarrow q) \wedge (q \Rightarrow p)$".

Basic propositions are taken to have a *truth-value* that is either *true* (T) or *false* (F). We do not allow "*maybe*" as a truth-value. The idea is that although we might not be able to practically determine if a proposition is true or false, if we had enough information, we could in principle determine if the statement is true or false.

2.1.2 *Precedence*

In ordinary algebra and arithmetic, if we write "$3 + 4 \times 5$", then we intend to compute "4×5" first, instead of "$3 + 4$". We could make this clear by means of parentheses ("$3 + (4 \times 5)$"), but it is usually understood that multiplication takes precedence over addition. In a similar way, "$\neg$" ("not") takes precedence over "$\wedge$" ("and"), which takes precedence over "$\vee$" ("or") which takes precedence over "$\Rightarrow$" ("implies") which takes precedence over "$\Leftrightarrow$" ("equivalent to"). Thus "$\neg\neg p \Rightarrow q \vee r$" means "$(\neg(\neg p)) \Rightarrow (q \vee r)$" . If the meaning is unclear, then you should insert parentheses wherever appropriate. For example, "$p \Rightarrow q \Rightarrow r$" is not meaningful as it could represent "$(p \Rightarrow q) \Rightarrow r$" or "$p \Rightarrow (q \Rightarrow r)$", which are quite different propositions. You do not need to worry about "$p \wedge q \wedge r$" since both ways of parenthesizing the expression ("$(p \wedge q) \wedge r$" and "$p \wedge (q \wedge r)$") are equivalent by the associative property (see Section 2.1.8). Sometimes you should parenthesize for ease of reading, even if the meaning is clear from the precedence rules. For example, it is better to write "$(p \wedge q \wedge r) \vee (a \wedge b \wedge c)$" than "$p \wedge q \wedge r \vee a \wedge b \wedge c$".

2.1.3   *Truth tables*

We can define the meaning of the operations in terms of truth tables, as shown in Figure 2.1.

p	$\neg p$
F	T
T	F

p	q	$p \wedge q$	$p \vee q$	$p \Rightarrow q$
F	F	F	F	T
F	T	F	T	T
T	F	F	T	F
T	T	T	T	T

Fig. 2.1: Truth tables for "not" ($\neg$), "and" ($\wedge$), "or" ($\vee$), and "implies" ($\Rightarrow$)

2.1.4   *Tautologies*

Tautologies are propositions that must be true simply because of their structure. For example, *"Today is hot or today is not hot."* Or, *"If today is hot, then today is hot or tomorrow will be cold."* These are clearly true regardless of the weather today or tomorrow. We can analyze these using propositional calculus. Let p represent the proposition *"Today is hot,"* and q represent the proposition *"Tomorrow will be cold."* Then the first tautology *"Today is hot or today is not hot "* is represented by the formula $p \vee \neg p$; the second tautology *"If today is hot, then today is hot or tomorrow will be cold "* is represented by $p \Rightarrow (p \vee q)$.

   We can look at the truth tables of these formulas:

p	$\neg p$	$p \vee \neg p$
F	T	T
T	F	T

No matter the truth value of p, $p \vee \neg p$ is always true. This makes $p \vee \neg p$ a tautology.

Now let us look at $p \Rightarrow (p \vee q)$.

p	q	$p \vee q$	$p \Rightarrow (p \vee q)$
F	F	F	T
F	T	T	T
T	F	T	T
T	T	T	T

The truth value of $p \Rightarrow (p \vee q)$ is T no matter the truth values of p and q. This makes $p \Rightarrow (p \vee q)$ a tautology.

2.1.5 *Rules of inference*

We can check if a proposition is a tautology by looking at its truth table. But not all true statements are tautologies: "For all real x, $x^2 \geq 0$." We need to move from truth tables to proving statements. This means we need *rules of inference*. Rules of inference are ways of combining previous true statements to produce new true statements that are logically and rigorously justified.

A proof can then be defined as a sequence of statements where each statement is either an axiom, or obtained from previous statements by a rule of inference.

The basic rule of inference is *modus ponens* or *MP*: from p and $p \Rightarrow q$ we can infer q. We can write this rule as

$$\frac{\begin{array}{c} p \\ p \Rightarrow q \end{array}}{\therefore q}$$

Many other rules of inference have been devised, but modus ponens is the only one that we really need. For example, $(p \wedge q) \Rightarrow p$ is a tautology, so we have the rule of inference

$$\frac{p \wedge q}{\therefore p}$$

We have simply skipped the step of writing down the tautology $(p \wedge q) \Rightarrow p$. This rule of inference is called *conjunction elimination*. A number of other rules of inference are given below that are combinations of modus ponens with appropriate tautologies.

A suitable set of axioms was devised by Jan Łukasiewicz: any formula obtained by substituting propositional expressions for a, b and c in the propositional expressions below is an axiom.

$$a \Rightarrow (b \Rightarrow a), \tag{2.1}$$

$$(a \Rightarrow (b \Rightarrow c)) \Rightarrow ((a \Rightarrow b) \Rightarrow (b \Rightarrow c)), \quad \text{and} \tag{2.2}$$

$$(\neg a \Rightarrow \neg b) \Rightarrow (b \Rightarrow a). \tag{2.3}$$

For example, if we substitute $p \wedge q$ for a and $p \vee q$ for b in (2.1) we get

$$(p \wedge q) \Rightarrow ((p \vee q) \Rightarrow (p \wedge q)).$$

If we define $a \vee b$ as $\neg a \Rightarrow b$ and define $a \wedge b$ as $\neg(a \Rightarrow \neg b)$, then the axioms of Łukasiewicz together with Modus Ponens can prove all tautologies, and only tautologies.

2.1.6 *Implication via assumption**

A strategy to show $p \Rightarrow q$ is to assume p and then show q. That is, we assume p is an axiom, and then write a proof of q using p. This strategy is called *implication via assumption*. Any proof of q with p as an axiom can be turned into a proof that $p \Rightarrow q$ without p as an axiom. Every statement φ in the proof with p as an axiom, is replaced with the statement $p \Rightarrow \varphi$ in the proof without p as an axiom. Here we outline the idea of how to transform such a proof of q assuming p into a proof of $p \Rightarrow q$. A complete formal proof should be carried out using induction on the number of lines of the proof (see Section 2.2.6). Note that $(p \Rightarrow r) \Rightarrow (((p \Rightarrow (r \Rightarrow s)) \Rightarrow (p \Rightarrow s))$ is a tautology.

proof assuming p	proof not assuming p	
r	$p \Rightarrow r$	(given)
$r \Rightarrow s$	$p \Rightarrow (r \Rightarrow s)$	(given)
	$(p \Rightarrow r) \Rightarrow ((p \Rightarrow (r \Rightarrow s)) \Rightarrow (p \Rightarrow s))$	
	$(p \Rightarrow (r \Rightarrow s)) \Rightarrow (p \Rightarrow s)$ (modus ponens)	
$\therefore s$	$\therefore p \Rightarrow s$	(modus ponens again)

2.1.7 *Inference steps*

Proof by contradiction is based on the fact that $(p \Rightarrow q) \Leftrightarrow (\neg q \Rightarrow \neg p)$ is a tautology. So to prove $p \Rightarrow q$ it is sufficient to show that $\neg q \Rightarrow \neg p$. We prove this by assuming $\neg q$ (denying the conclusion) to show $\neg p$ (the hypothesis is contradicted). Note that $\neg q \Rightarrow \neg p$ is the *contrapositive* of $p \Rightarrow q$.

Given p (or q) we can infer $p \vee q$ since $p \Rightarrow (p \vee q)$ is a tautology. This is called *disjunction introduction*.

Given p and q we can infer $p \wedge q$ since $p \Rightarrow (q \Rightarrow (p \wedge q))$ is a tautology. This is called *conjunction introduction*.

Given $p \wedge q$ we can infer p (and q) since $p \wedge q \Rightarrow p$ and $p \wedge q \Rightarrow q$ are tautologies. These are called *conjunction elimination*.

Separation of cases is based on the fact that given $p \Rightarrow r$ and $q \Rightarrow r$ we can infer $p \vee q \Rightarrow r$ since $(p \Rightarrow r) \wedge (q \Rightarrow r) \Rightarrow (p \vee q \Rightarrow r)$ is a tautology.

From $\neg\neg p$ we can infer p since $\neg\neg p \Rightarrow p$ is a tautology.

For counterexamples, if we can show $p \wedge \neg q$, then since $\neg(p \Rightarrow q) \Leftrightarrow \neg(\neg p \vee q) \Leftrightarrow \neg\neg p \wedge \neg q \Leftrightarrow p \wedge \neg q$ we can infer $\neg(p \Rightarrow q)$.

2.1.8 *The algebra of propositions*

The main rules of propositions are listed below:

- *Associativity*:

$$(p \wedge q) \wedge r \Leftrightarrow p \wedge (q \wedge r), \tag{2.4}$$

$$(p \vee q) \vee r \Leftrightarrow p \vee (q \vee r). \tag{2.5}$$

This means you do not need to worry about parenthesizing $p \wedge q \wedge r \wedge s \wedge \cdots$ or $p \vee q \vee r \vee s \vee \cdots$.

- *Commutativity*:

$$p \wedge q \Leftrightarrow q \wedge p, \tag{2.6}$$

$$p \vee q \Leftrightarrow q \vee p. \tag{2.7}$$

- *DeMorgan's laws*:

$$\neg(p \wedge q) \Leftrightarrow \neg p \vee \neg q, \tag{2.8}$$

$$\neg(p \vee q) \Leftrightarrow \neg p \wedge \neg q. \tag{2.9}$$

These are actually very helpful in programming:

$$\neg(x > 0 \wedge y > 0) \Leftrightarrow \neg(x > 0) \vee \neg(y > 0) \Leftrightarrow (x \leq 0 \vee y \leq 0)$$

so

```
if ( x > 0 and y > 0 )
{  . . . . . .  }
else /* x ≤ 0 or y ≤ 0 */
{  . . . . . .  }
```

- *Double negation*:

$$\neg\neg p \Leftrightarrow p. \tag{2.10}$$

This is analogous to "$-(-x) = x$" in ordinary arithmetic.

- *Disjunction with negation*:

$$p \vee \neg p.$$

This is also known as the "*law of the excluded middle*". A proposition is either true or false; "maybe" is not allowed.

- *Distributive laws*:

$$p \wedge (q \vee r) \Leftrightarrow (p \wedge q) \vee (p \wedge r), \tag{2.11}$$

$$p \vee (q \wedge r) \Leftrightarrow (p \vee q) \wedge (p \vee r). \tag{2.12}$$

In ordinary arithmetic $a \times (b + c) = a \times b + a \times c$, but almost always $a + (b \times c) \neq (a + b) \times (a + c)$. However, in the algebra of propositions, both $\wedge$ and $\vee$ distributes over the other operation.

- *Definition of "implies" and "equivalent to"*:

$$p \Rightarrow q \Leftrightarrow (\neg p) \vee q, \tag{2.13}$$

$$p \Leftrightarrow q \Leftrightarrow (p \Rightarrow q) \wedge (q \Rightarrow p). \tag{2.14}$$

Every expression in the propositional calculus can be reduced to an expression with only "$\wedge$", "$\vee$" and "$\neg$".

2.1.9 *Boolean algebra*

A *Boolean algebra* is a set together with operations $\wedge$, $\vee$ and $\neg$ satisfying the axioms (2.4–2.12) with "$=$" taking the place of "$\Leftrightarrow$", which also has two elements denoted 0 (zero) and 1 (one) where

$$1 \wedge p = p \qquad \text{for all } p,$$

$$0 \vee p = p \qquad \text{for all } p.$$

A standard example of a Boolean algebra is the collection of subsets of a given set S. We interpret "$\wedge$" as intersection: for $A, B \subseteq S$ we let $A \wedge B = A \cap B = \{ x \in S \mid x \in A \text{ and } x \in B \}$; we interpret "$\vee$" as union: $A \vee B = A \cup B = \{ x \in S \mid x \in A \text{ or } x \in B \}$; finally, we interpret "$\neg$" as the complement: $\neg A = S \backslash A = \{ x \in S \mid x \notin A \}$. For this interpretation, we set $1 = S$ and $0 = \emptyset$, the empty set.

2.2 Expressions, predicates, and quantifiers

In mathematics we need to deal with objects like numbers as well as propositions. However, to connect these with logic, we need to have statements about these objects like "$x = 7$" or "$y > 2$" or "$x^2 + y^2 > 0$". For this we need *expressions* and *predicates*. In any mathematical theory there must be *constants*, *variables*, and *operations*. In arithmetic, constants can be named numbers like 7, -33, 2567.0126 or π. Variables simply use a name, like "x" or "w" or even "ζ" (zeta) in place of a number. To constants and variables we can apply operations such as negation, addition, subtraction, multiplication, division (provided the denominator is not zero), and exponentiation. Furthermore, we can apply these operations to the expressions we have created to create even more expressions. There is no limit to the complexity of the expressions we can build in this way.

Logical propositions about constants, variables and expressions are called *predicates*. For example, "$x > y$" is a proposition about variables x and y; "n is prime" is a proposition about n. Whether the proposition is true or not depends on the value of the variables. We can think of a proposition as a function from the set of possible values of an expression to a truth-value, which must be in the set $\{F, T\}$. Often we write a predicate as a function $P(x)$ or $R(x, y)$. Since the value of a predicate is a truth-value, we can combine them with all the usual operations in propositional calculus: $\wedge$, $\vee$, $\neg$, $\Rightarrow$ and $\Leftrightarrow$. This gives statements in *predicate calculus*. However, if we have any variables in the statement, we cannot usually tell if the statement is true unless we specify the values of all the variables. Instead, we usually want to know if the statement is true for all values of the variables, or if it is true for at least one value of the variables. To express this we need *quantifiers*. Usually there is a common understanding of what values the variables can range over. They might range over whole numbers, real numbers, people, vectors or points in space, so some other collection of objects. Whatever set the variables can range over is called the *universe*, at least for the problem at hand.

The two quantifiers that are used all the time are:

- "For all": $\forall$
  "$\forall x (x^2 \geq 0)$" means "for every x, $x^2 \geq 0$."
- "There exists": $\exists$
  "$\exists x (x^2 > 0)$" means "there exists x where $x^2 > 0$."

We can combine quantifiers for different variables, as long as the variable associated with the quantifier is *free* (that is, does not already have a quantifier) in a logical formula. A variable that is not free in a logical formula is called a *bound variable.* For example, in the formula $x > 0 \Rightarrow \exists y(0 < y < x)$, x is a free variable but y is a bound variable.

There is a restriction on the variables we can use in a quantifier $\forall$ or $\exists$: for a logical formula $\forall z\,\phi$ or $\exists z\,\phi$, then the variable z must be *free in* ϕ. That is, z must not be associated with (or *bound by*) a quantifier. The restriction that in a quantified formula $\forall z\,\phi$ or $\exists z\,\phi$ that z must be a free variable in ϕ (or at least, not a bound variable in ϕ) means that $\forall x\,(x > 0 \Rightarrow \exists y(0 < y < x))$ is a formula in predicate calculus ("for every positive x there is a positive y where $y < x$") which may be true or false[1], but $\forall y\,(x > 0 \Rightarrow \exists y(0 < y < x))$ is not.

To express the idea that every positive number y has a positive square root x, we can use quantifiers:

$$\forall y(y > 0 \Rightarrow \exists x(x > 0 \wedge x^2 = y)).$$

Pulling this statement apart, it says:

For all y, if y is positive then there is an x which is positive and $x^2 = y$.

Some mathematicians use an additional predicate "$\exists!$" to represent "there exists exactly one ..." For the statement "every positive number y has exactly one positive square root x" could be expressed by

$$\forall y(y > 0 \Rightarrow \exists! x(x > 0 \wedge x^2 = y)).$$

This does not really add a lot to the language because we can express "$\exists!$" in terms of the other predicates and some extra logic:

$$\exists! x\, P(x) \;\Leftrightarrow\; \exists x(P(x) \wedge \forall y\,(P(y) \Rightarrow y = x)).$$

In English, $\exists x(P(x) \wedge \forall y\,(P(y) \Rightarrow y = x))$ could be expressed as

there is an x where $P(x)$ is true, and for any y where $P(y)$ is true then $y = x$.

A convention in logic is that if we write down a statement that has free variables, we mean that the statement is true if it is true for all possible values (and combination of values) of the variables. So

$$x^2 + y^2 \geq 0$$

is a true statement over the real numbers, but

$$x^2 - y^2 > 0$$

---

[1] In this case, the statement is true if the universe is the set of real numbers or the set of rational numbers, but not if the universe is the set of whole numbers.

is not, even though there are *some* values of x and y that can make $x^2 - y^2 > 0$ true.

Sometimes we have to negate quantified statements: *"Not every person is Aristotle."* This can be written in predicate calculus over the universe of people as "$\neg \forall x(x = \text{Aristotle})$". If not every person is Aristotle, then there must be a person who is not Aristotle: "$\exists x(x \neq \text{Aristotle})$" or *"There is a person who is not Aristotle."* This is an example of the general rules:

$$\neg \forall x\, P(x) \Leftrightarrow \exists x\, \neg P(x), \tag{2.15}$$

$$\neg \exists x\, P(x) \Leftrightarrow \forall x\, \neg P(x). \tag{2.16}$$

2.3 Rules of inference

2.3.1 *Rules of inference for propositional calculus*

We have seen rules of inference for propositional calculus: if we have propositions p and $p \Rightarrow q$ then we can infer q. Any tautology can be treated as an axiom of propositional calculus.

In predicate calculus, we can treat as an axiom any tautology where each proposition in the tautology is replaced by a formula of predicate calculus. For example, $p \Rightarrow p \vee q$ is a tautology of propositional calculus. We can substitute "$x > y$" for p and "$y^2 < y$" for q. This gives the tautology of predicate calculus

$$(x > y) \Rightarrow ((x > y) \vee (y^2 < y)).$$

Or we could substitute "$\exists y\,(y < x)$" for p and "$\forall x(x > 0 \Rightarrow x > y^2)$" for q giving a different tautology of predicate calculus:

$$\exists y\,(y < x) \Rightarrow (\exists y\,(y < x) \vee \forall x(x > 0 \Rightarrow x > y^2)).$$

In predicate calculus, just as in propositional calculus, if we have proven the logical formulas (or propositions) p and $p \Rightarrow q$, then we can infer q. For example, if we have proven that

$$x \neq 0 \qquad \text{and}$$

$$x \neq 0 \Rightarrow x^2 > 0$$

then we can infer that $x^2 > 0$.

2.3.2 *Rules of inference with quantifiers*

There are additional rules of inference that apply regarding quantifiers. We use $\phi(x)$ to denote a formula with x as a free variable; then for an expression e we use $\phi(e)$ to denote the formula with all (free) occurrences of x replaced by e.

The first rule of inference is that

$$\forall x \, \phi(x)$$

infer $\phi(e)$

provided e does not contain any bound variables of $\phi(x)$. We need this restriction on e: if $\phi(x)$ is $\exists y(y > x)$ and we allowed the expression $e = y$ then from $\forall x \, \exists y(y > x)$ (which is true for real numbers) we could infer $\exists y(y > y)$, which is clearly false. But $e = y$ is disallowed because y is bound by $\exists$ in $\exists y(y > x)$.

Since a statement with a free variable is taken to mean that the statement is true for any value of that free variable, we have the inference

$$\phi(x)$$

infer $\forall x \, \phi(x)$.

Combining the previous two inference rules, if e is an expression, we have the inference

$$\phi(x)$$

infer $\phi(e)$.

For existential quantifiers, the rules of inference are a little more complex:

$$\exists x \, \phi(x, y, z, \ldots)$$

infer $\phi(f(y, z, \ldots))$,

where f is a *new* function symbol and y, z, $\ldots$ are the other free variables in ϕ. On the other hand, if e is any constant expression (that is, contains no variables but only constants) then

$$\phi(e)$$

infer $\exists x \, \phi(x)$.

The reason for allowing only constant expressions is that in an *empty* universe $\exists x \, \phi(x)$ is *never* true, since nothing exists. (See Exercise 2.24.) As soon as something exists, then we have the implication

$$\forall x \, \phi(x) \;\Rightarrow\; \exists x \, \phi(x).$$

2.4   Axioms of equality and inequality

The most common predicate is "equals" ("$=$"). In spite of using it all the time, we often forget that it is a predicate and it has its own axioms:

- *Symmetry*: if $x = y$ then $y = x$. ($x = y \implies y = x$)
- *Transitivity*: if $x = y$ and $y = z$ then $x = z$. ($x = y \wedge y = z \implies x = z$)
- *Reflexivity*: for any x, $x = x$. ($\forall x \, (x = x)$).

Predicates of two variables (which are also called *relations*) that have these three properties of symmetry, transitivity and reflexivity are called *equivalence relations*.

There is another collection of axioms that express the idea that *if two things are equal, then one can be substituted for the other*. So if we have a function or operation $f(x_1, x_2, \ldots, x_m)$ we have the axiom

$$x_1 = y_1 \wedge x_2 = y_2 \wedge \cdots \wedge x_m = y_m \implies$$
$$f(x_1, x_2, \ldots, x_m) = f(y_1, y_2, \ldots, y_m).$$

We do not often need to use the axioms of equality explicitly.

Inequalities such as "less than" are normally applied just to real numbers, but sometimes we want to use inequalities in a more general sense, or define "less than" in some other context. Here we will use the symbol "$\preceq$" to mean some inequality predicate that is not necessarily the usual inequality for numbers. The following assumptions are basic to any "inequality":

- *Antisymmetric*: $a \preceq b$ and $b \preceq a$ implies $a = b$.
- *Transitive*: $a \preceq b$ and $b \preceq c$ implies $a \preceq c$.

Any predicate that is both antisymmetric and transitive is a *partial order*. We have the corresponding *strict* inequality: we define "$a \prec b$" to mean "$a \preceq b$ and $a \neq b$". The strict inequality is also antisymmetric and transitive ($a \prec b$ and $b \prec a$ are never both true, even if $a = b$). An important property of the standard inequality of numbers is that every pair of numbers is comparable. This makes the ordinary "less than" predicate for numbers a *total order*:

- *Total order*: for all a, b, we have either $a \prec b$, $b \prec a$, or $a = b$.

Using antisymmetry is a standard technique for proving equality in many different partial (and total) orders.

2.5 Dealing with sets

Sets are collections of objects, called elements. In fact, sets are nothing but the collection of their elements.

The central predicate for sets is "$\in$": "$x \in A$" means "x is an element of the set A". This predicate defines sets:

$$\text{``}A = B\text{''} \text{ means ``}x \in A \iff x \in B.\text{''}$$

That is, a set is defined by what it contains.

2.5.1 *Set operations*

The basic set operations are

$$A \cap B = \{\, x \mid x \in A \,\wedge\, x \in B \,\} \qquad \text{(intersection)},$$
$$A \cup B = \{\, x \mid x \in A \,\vee\, x \in B \,\} \qquad \text{(union)},$$
$$A \backslash B = \{\, x \mid x \in A \,\wedge\, x \notin B \,\} \qquad \text{(set difference)}.$$

The algebra of intersections and unions of sets is like the Boolean algebra of $\wedge$ and $\vee$: $A \cap (B \cap C) = (A \cap B) \cap C$, $A \cap (B \cup C) = (A \cap B) \cup (A \cap C)$, etc. Set differences are not quite like negation in propositional logic, but they are closely related. For example, from the distributive and De Morgan's laws it can be shown that $(A \backslash B) \backslash C = A \backslash (B \cup C)$, but $(A \backslash B) \cup C = (A \cup C) \backslash (B \backslash C)$:

$$\begin{aligned}
(A \backslash B) \cup C &= \{\, x \mid (x \in A \wedge x \notin B) \vee x \in C \,\} \\
&= \{\, x \mid (x \in A \vee x \in C) \wedge (x \notin B \vee x \in C) \,\} \\
&= \{\, x \mid (x \in A \vee x \in C) \wedge \neg(x \in B \wedge x \notin C) \,\} \\
&= (A \cup C) \backslash (B \backslash C).
\end{aligned}$$

The predicate "$\subseteq$" on sets is a partial order: "$A \subseteq B$" means "for all x, $x \in A$ implies $x \in B$". Note that the antisymmetric property means that "$A \subseteq B$ and $B \subseteq A$ implies $A = B$". This is a common way to show that two sets are equal: first show that one set is a subset of the other, then show the reverse inclusion.

The next operation on sets is the *power set* operation:

$$\mathcal{P}(S) = \{\, A \mid A \subseteq S \,\},$$

the set of subsets of S. For example,

$$\mathcal{P}(\{1, 2, 3\}) = \{\, \emptyset,\, \{1\},\, \{2\},\, \{3\},\, \{1, 2\},\, \{1, 3\},\, \{2, 3\},\, \{1, 2, 3\} \,\}.$$

A special set is the *empty set*: $x \notin \emptyset$ for all x. Starting with the empty set, unions, intersections and power set operations it is possible to build up an entire universe of sets. For example, $\mathcal{P}(\emptyset) = \{\emptyset\}$, $\mathcal{P}(\{\emptyset\}) = \{\emptyset, \{\emptyset\}\}$, and $\mathcal{P}(\mathcal{P}(\{\emptyset\})) = \{\emptyset, \{\emptyset\}, \{\{\emptyset\}\}, \{\emptyset, \{\emptyset\}\}\}$.

Given a set S and a predicate with one variable $P(x)$ there is a set $\{\, x \in S \mid P(x) \,\}$ of everything in S that satisfies P. Given a set J and a function $\alpha \mapsto A_\alpha$ where A_α is a set, then the union

$$\bigcup_{\alpha \in J} A_\alpha = \{\, x \mid \exists \beta\, [\beta \in J \wedge x \in A_\beta] \,\}$$

is also set. Combining these two facts means that the intersection

$$\bigcap_{\alpha \in J} A_\alpha = \left\{ x \in \bigcup_{\alpha \in J} A_\alpha \mid \forall \beta \, (\beta \in J \Rightarrow x \in A_\beta) \right\}$$

is also a set.

But some things are *not* sets, even though they might look like sets. A famous example was devised by the mathematical philosopher Bertrand Russell:

$$R = \{ x \mid x \notin x \}. \tag{2.17}$$

This is an example of a *self-defeating object* — its very definition implies it cannot exist: if $R \notin R$ then by definition of R we have $R \in R$; on the other hand, if $R \in R$ then by definition of R we have $R \notin R$. So $R \in R$ is true if and only if $R \in R$ is false, which is impossible.

A consequence of Russell's famous "set" is that there is no universal set, for if there were a set U of everything, then we could write $R = \{ x \in U \mid x \notin x \}$. But we already know that R does not exist.

2.5.2 *Special sets*

There are a number of special sets apart from the empty set:

> $\mathbb{N}$ the set of natural numbers $\{1, 2, 3, \ldots\}$,
>
> $\mathbb{Z}$ the set of integers $\{\ldots, -4, -3, -2, -1, 0, 1, 2, 3, 4, \ldots\}$,
>
> $\mathbb{Q}$ the set of rational numbers $\{ a/b \mid a, b \in \mathbb{Z}, \, b \neq 0 \}$,
>
> $\mathbb{R}$ the set of real numbers,
>
> $\mathbb{C}$ the set of complex numbers.

These are all infinite sets. Usually we begin with the natural numbers $\mathbb{N}$, and from this define integers $\mathbb{Z}$, then rationals $\mathbb{Q}$, then real numbers $\mathbb{R}$, and finally complex numbers $\mathbb{C}$. Each set in this hierarchy is built out of the previous set: $\mathbb{Z}$ is made of differences of natural numbers ($\mathbb{N}$); $\mathbb{Q}$ is made of quotients of integers ($\mathbb{Z}$); $\mathbb{R}$ is made of limits of convergent sequences of rational numbers ($\mathbb{Q}$); $\mathbb{C}$ is made of sums $x + \mathrm{i}\, y$ where x, y are real numbers ($\mathbb{R}$).

2.5.3 *Functions*

A *function* $f \colon A \to B$ is a rule or map from a set A to a set B where for any $a \in A$, $f(a) \in B$. We say that $A = \mathrm{dom}(f)$ is the domain of the function f; the range of f is the set $\mathrm{range}(f) = \{ f(a) \mid a \in A \}$. Any rule can be used to define a function, as long as the output $f(a)$ is uniquely specified for any given input

$a \in A$. We can represent a function as a set as follows. Given two values $a \in A$ and $b \in B$ there is an ordered pair (a, b). The crucial property of the ordered pair (a, b) is that $(a, b) = (c, d)$ if and only if $a = c$ and $b = d$. One way to ensure this is if we define $(a, b) = \{\{a\}, \{a, b\}\}$. There are other ways of defining ordered pairs that guarantees the crucial property $(a, b) = (c, d) \Rightarrow a = c \land b = d$, but we will use this one. The set $A \times B = \{(a, b) \mid a \in A, \ b \in B\}$ of ordered pairs is called the *Cartesian product* of A and B.

The function $f \colon A \to B$ can be represented by its graph: $\mathrm{graph}(f) = \{(a, f(a)) \mid a \in A\} \subseteq A \times B$. We say that two functions $f, g \colon A \to B$ are equal if and only if $f(a) = g(a)$ for all $a \in A$. That is, two functions are equal if they always give the same output for the same input. This is equivalent to $\mathrm{graph}(f) = \mathrm{graph}(g)$.

A function $f \colon A \to B$ is *one-to-one* if $f(a_1) = f(a_2) \Rightarrow a_1 = a_2$; it is *onto* if for every $b \in B$ there is an $a \in A$ where $f(a) = b$. A function that is both one-to-one and onto is called a *bijection* or a *one-to-one correspondence*. If $f \colon A \to B$ is a bijection, then there is an *inverse function* $g \colon B \to A$: $g(f(a)) = a$ for all $a \in A$ and $f(g(b)) = b$ for all $b \in B$. We write $g = f^{-1}$ for the inverse function of f. The *composition* of two functions $f \colon A \to B$ and $g \colon B \to C$ is the function $g \circ f \colon A \to C$ where $g \circ f(a) = g(f(a))$. The composition of one-to-one functions is one-to-one; the composition of onto functions is onto. Thus the composition of bijections is also a bijection, and $(g \circ f)^{-1} = f^{-1} \circ g^{-1}$. Composition of functions is an associative operation in the sense that if $f \colon A \to B$, $g \colon B \to C$ and $h \colon C \to D$, we have $(h \circ g) \circ f = h \circ (g \circ f)$. To see this, note that for all $a \in A$,

$$(h \circ g) \circ f(a) = (h \circ g)(f(a)) = h(g(f(a))) = h((g \circ f)(a)) = h \circ (g \circ f)(a).$$

Thus $(h \circ g) \circ f = h \circ (g \circ f)$ as functions $A \to D$. This means we do not need to worry about how we parenthesize compositions of functions. If $f \colon A \to A$ then we can compose f with itself: $f \circ f \colon A \to A$, $f \circ f \circ f \colon A \to A$, etc. In general we can define $f^2 = f \circ f$, $f^3 = f \circ f \circ f$, etc. But this f^2 is quite different from the function $g(a) = f(a)^2$ if the values $f(a)$ are numbers. For example if $f \colon \mathbb{R} \to \mathbb{R}$ is given by $f(x) = 2x + 3$, then $f^2(x) = f(f(x)) = 2f(x) + 3 = 2(2x + 3) + 3 = 4x + 9$, which is quite different from $g(x) = f(x)^2 = (2x + 3)^2 = 4x^2 + 12x + 9$.

2.6 Proof by induction

Making chains of dominoes fall over is lots of fun. The principle is that if we can knock domino n fall over, then it will knock over domino $n + 1$ which will fall over, as illustrated in Figure 2.2. If we knock over the first domino, then the second will fall, so will the third, and the fourth, fifth, etc.

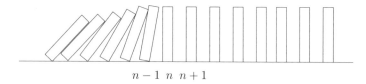

$$n - 1 \quad n \quad n + 1$$

Fig. 2.2: Dominoes

We can represent this with $P(n)$ the predicate "the nth domino will fall over". Then, if the universe is taken to be the set of natural numbers $\mathbb{N} = \{\, 1, 2, 3, \ldots \,\}$, the principle of mathematical induction is

$$(P(1) \, \wedge \, \forall n(P(n) \Rightarrow P(n + 1))) \; \Rightarrow \; \forall n\, P(n). \tag{2.18}$$

In fact, with $\mathbb{N}$ as the universe, we take (2.18) as an axiom for *any* predicate P. These are the axioms of mathematical induction. A proof by induction has the following parts:

- *Base step*: We prove that $P(1)$ holds. This is the starting point for the induction argument. We do not need to start with $n = 1$; often it is convenient to start with $n = 0$. Usually (but not always) this is quite easy.
- *Induction step*: We prove that $\forall n(P(n) \Rightarrow P(n + 1))$. This is where most of the work is.
- *Conclusion*: Having proven $P(1)$ and $\forall n(P(n) \Rightarrow P(n + 1))$ we can conclude that $\forall n\, P(n)$.

When we are carrying out the induction step, we can assume $P(n)$, which is called the *induction hypothesis*, in order to prove $P(n + 1)$. Often proofs will have a statement with something like "*...by the induction hypothesis...*" where the assumption of $P(n)$ is actually used.

There are variations on this: We can start with proving $P(k)$ for some k, then prove $\forall n\, (n \geq k \wedge P(n) \Rightarrow P(n + 1))$, in order to conclude that $\forall n\, (n \geq k \Rightarrow P(n))$; that is, concluding that $P(n)$ is true for all $n \geq k$.

Another variation is called *complete induction*: The base case $P(1)$ also holds, but instead of assuming only that $P(n)$ is true to show that $P(n + 1)$ is true, we can assume that $P(r)$ is true for all $1 \leq r \leq n$ in order to show that $P(n + 1)$ is true.

Induction is not only used to prove statements, it can also be used to define certain concepts. For example, the sum

$$\sum_{k=a}^{b} u_k = u_a + u_{a+1} + \cdots + u_b \tag{2.19}$$

can be defined by first defining it for $b = a$, and then using $\sum_{k=a}^{b} u_k$ to define $\sum_{k=a}^{b+1} u_k$:

$$\sum_{k=a}^{a} u_k = u_a, \tag{2.20}$$

$$\sum_{k=a}^{b+1} u_k = \left[\sum_{k=a}^{b} u_k \right] + u_{b+1}. \tag{2.21}$$

Now let us get some practice at mathematical induction.

2.6.1 *Some sums*

The following is a well-known result.

Theorem 2.1. $\displaystyle\sum_{k=1}^{n} k = \frac{1}{2}n(n+1)$ *for* $n = 1, 2, 3, \ldots$.

Proof. Let $P(n)$ be the statement "$\sum_{k=1}^{n} k = \frac{1}{2}n(n+1)$". We prove this by induction on n.

> *It is best to declare our intention to prove the statement by induction first.*

Base case: Note that $P(1)$ is the statement that $1 = \sum_{k=1}^{1} k = \frac{1}{2} 1 \times (1+1) = 1$, which is true.

> *The base case is often easy.*

Induction step: Suppose true for $n = m$; show true for $n = m + 1$:

> *We now show $\forall m \, (P(m) \Rightarrow P(m+1))$.*

Suppose $P(m)$. That is, $\sum_{k=1}^{m} k = \frac{1}{2}m(m+1)$. We want to show $P(m+1)$; that is, $\sum_{k=1}^{m+1} k = \frac{1}{2}(m+1)((m+1)+1)$.

> *This unwraps $P(m)$ and $P(m+1)$.*

Now we use the inductive definition of $\sum$:

$$\sum_{k=1}^{m+1} k = \left[\sum_{k=1}^{m} k \right] + (m+1)$$

$$= \frac{1}{2}m(m+1) + (m+1) \qquad (\text{from } P(m))$$

$$= (m+1)\left[\frac{1}{2}m + 1 \right] = \frac{1}{2}(m+1)(m+2)$$

$$= \frac{1}{2}(m+1)((m+1)+1),$$

as we wanted.

Conclusion: $P(n)$ holds for $n = 1, 2, 3, \ldots$; that is, $\sum_{k=1}^{n} k = \frac{1}{2}n(n+1)$ for $n = 1, 2, 3, \ldots$.

$\square$

There are many similar proofs to this of formulas for $\sum_{k=1}^{n} k^r$ where r is a positive integer. See the Exercises for more examples.

2.6.2 *Fibonacci numbers*

Fibonacci numbers are a well-known sequence of integers that is generated by a recurrence:

$$F_{n+1} = F_n + F_{n-1}, \qquad \text{for } n = 2, 3, 4, \ldots \text{ and} \qquad (2.22)$$
$$F_1 = 1, \qquad F_2 = 1. \qquad (2.23)$$

That is, each F_n is the sum of the previous two entries in the sequence. This gives the sequence

$$1, \; 1, \; 2, \; 3, \; 5, \; 8, \; 13, \; 21, \; 34, \; 55, \; 89, \; 144, \; 233, \; 377, \; \ldots$$

Most basic properties of Fibonacci numbers are obtained through proofs by induction. Here are two easy examples:

Theorem 2.2. *For $n \geq 1$, $\displaystyle\sum_{i=1}^{n} F_i = F_{n+2} - 1$.*

Proof. We prove $\displaystyle\sum_{i=1}^{n} F_i = F_{n+2} - 1$ by induction on n.

Base case: For $n = 1$ we have $F_1 = 1 = 2 - 1 = F_3 - 1$, which is true.

Induction step: Suppose true for $n = m$; show true for $n = m + 1$: So we suppose that

$$\sum_{i=1}^{m} F_i = F_{m+2} - 1,$$

as it is the induction hypothesis. Then

$$\sum_{i=1}^{m+1} F_i = \left[\sum_{i=1}^{m} F_i \right] + F_{m+1}$$

$$= F_{m+2} + F_{m+1} = F_{m+3} = F_{(m+1)+2}$$

so the statement is true for $n = m + 1$.

Conclusion: Thus $\sum_{i=1}^{n} F_i = F_{n+2} - 1$ is true for all integers $n \geq 1$. □

The next example is a little more complex, but not much.

Theorem 2.3. *For $n \geq 1$, $\sum_{i=1}^{n} F_i^2 = F_n F_{n+1}$.*

Proof. We prove $\sum_{i=1}^{n} F_i^2 = F_n F_{n+1}$ by induction on n.

Base case: For $n = 1$ the statement becomes $F_1^2 = F_1 F_2$; that is, $1^2 = 1 \times 1$, which is true.

Induction step: Suppose true for $n = m$; show true for $n = m + 1$: Suppose $\sum_{i=1}^{m} F_i^2 = F_m F_{m+1}$. Then

$$\sum_{i=1}^{m+1} F_i^2 = \left[\sum_{i=1}^{m} F_i^2 \right] + F_{m+1}^2$$

$$= F_m F_{m+1} + F_{m+1}^2 = F_{m+1}(F_m + F_{m+1}) = F_{m+1} F_{m+2},$$

so the statement is true for $n = m + 1$.

Conclusion: Thus $\sum_{i=1}^{n} F_i^2 = F_n F_{n+1}$ is true for all integers $n \geq 1$. □

2.6.3 *Egyptian fractions*

Ancient Egyptians apparently did not like having numerators other than one in their fractions, but with different denominators. For example, $2/3$ would be represented as $\dfrac{1}{2} + \dfrac{1}{6}$. The question is: Can all positive fractions less than one be represented in this way? The answer is yes, and can be shown by an application of complete induction.

Note that we prove a little more by induction than just

$$\frac{a}{b} = \frac{1}{r_1} + \frac{1}{r_2} + \cdots + \frac{1}{r_m}$$

with $0 < r_1 < r_2 < \cdots < r_m$. We get a lower bound on r_1: $r_1 \geq b/a$. We need this in order to show the inequalities hold after the induction step. After all, $a/b = 1/b + 1/b + \cdots + 1/b$ (with a terms) is *not* an Egyptian fraction. Sometimes, proving more is easier in induction proofs.

Theorem 2.4. *If a/b is a fraction less than one with integers a, $b > 0$, then there are integers $b/a \leq r_1 < r_2 < \cdots < r_m$ where*

$$\frac{a}{b} = \frac{1}{r_1} + \frac{1}{r_2} + \cdots + \frac{1}{r_m}.$$

Proof. We use complete induction on a to prove the statement.

Base case: $a = 1$. This case is easy: put $r_1 = b$ so $a/b = 1/b = 1/r_1$.

Induction step: Suppose true for $1 \leq a < c$. Show true for $a = c$. Now suppose $c/b < 1$, so $b/c > 1$. Either b and c have a common factor, or they do not.

Suppose $d > 1$ is a common factor of b and c then $c/b = (c/d)/(b/d)$. By the induction hypothesis, $(c/d)/(b/d)$ can be written as an Egyptian fraction, which completes the induction step in this case.

Now suppose that b and c have no common factor other than one. Let r_1 be the smallest integer so that $r_1 > b/c$. Clearly $r_1 > 1$. Then

$$0 < \frac{c}{b} - \frac{1}{r_1} = \frac{r_1 c - b}{b r_1}.$$

If $r_1 c - b \geq c$ then $(r_1 - 1)c - b \geq 0$. If $(r_1 - 1)c - b = 0$ then $c/b = 1/(r_1 - 1)$, and the induction step is completed for this case. However, if $(r_1 - 1)c - b > 0$, then $r_1 - 1 > b/c$, which contradicts the definition of r_1 as the smallest integer $> b/c$. So we cannot have $r_1 c - b \geq c$. That is, $r_1 c - b < c$. By the induction hypothesis, we can write $(r_1 c - b)/(b r_1)$ as an Egyptian fraction:

$$\frac{c}{b} - \frac{1}{r_1} = \frac{r_1 c - b}{b r_1} = \frac{1}{r_2} + \frac{1}{r_3} + \cdots + \frac{1}{r_m},$$

where $b r_1/(r_1 c - b) \leq r_2 < r_3 < \cdots < r_m$. Then

$$\frac{c}{b} = \frac{1}{r_1} + \frac{1}{r_2} + \frac{1}{r_3} + \cdots + \frac{1}{r_m},$$

where $b/c \leq r_1 < b r_1/c < b r_1/(r_1 c - b) \leq r_2 < r_3 < \cdots < r_m$ since $b/c > 1$ and $0 < r_1 c - b < c$.

Thus the result holds for $a = c$.

Conclusion: Therefore, for $a = 1, 2, 3, \ldots$ and b a positive integer, there are integers $b/a \leq r_1 < r_2 < \cdots < r_m$ where $a/b = 1/r_1 + 1/r_2 + \cdots + 1/r_m$. $\square$

2.7 Proofs and algorithms

Algorithms carry out tasks, but how do we know they perform the right task? We can look at someone's code and often it is clear. But not always. All too often, some detail has been overlooked which makes the code fail in some way. Testing is good, and uncovers many, if not most, errors. Proving that an algorithm is correct in a formal way can be an excellent way to ensure quality control, especially if the key parts of the proof are embedded in the code:

```
/* precondition: n > 0 */
s ← 0
for i ← 1, 2, ..., n
   s ← s + i
end
/* postcondition: s = n(n + 1)/2 */
```

Provided the *precondition* is true before the execution of the code fragment, the *postcondition* should be true at the end of the execution of the code fragment. These are examples of *assertions*: statements in an algorithm that should be true whenever execution of the algorithm is at that point. Often we will write "pre" instead of "precondition", and "post" for "postcondition" in the algorithm examples.

```
/* pre: n ≥ 1 */
s ← a₁
for i ← 2, ..., n
   s ← s + aᵢ
end
/* post: s = ∑ⁿᵢ₌₁ aᵢ */
```

Effectively using preconditions and postconditions involves not only checking that they are valid for small pieces of code, and also combining them to prove facts about larger pieces of code. For example:

```
/* pre:  p */
code_block_1;
```

```
/* post: q */
/* pre:  r */
code_block_2;
/* post: s */
```

If $q \Rightarrow r$, then as long as p holds in the first line then s holds in the last line, and we have valid preconditions and postconditions.

If/then/else structures can also be incorporated:

```
/* pre: p */
if ( q )
   /* pre: p ∧ q */
   code_block_1;
   /* post: r₁ */
else /* ¬q */
   /* pre: p ∧ (¬q) */
   code_block_2;
   /* post: r₂ */
end
/* post: r₁ ∨ r₂ */
```

As a simple example, finding the maximum of two numbers can be done like this:

```
if ( a ≥ b )
   /* pre: a ≥ b */
   m ← a
   /* post: m = a ∧ a ≥ b */
else /* ¬(a ≥ b) */
   /* pre: a < b */
   m ← b
   /* post: m = b ∧ b ≥ a */
end
/* post: m = max(a, b) since
   (m = a ∧ a ≥ b) ∨ (m = b ∧ b ≥ a) ⇒ (m = max(a, b)) */
```

Most codes do not need this level of documentation. However, pre- and post-conditions do make it easier to check.

To go further with this idea, we need to understand how to deal with loops. That needs induction. Consider the while loop:

```
/* pre (1): p */
```

```
while ( q )
  /* pre (2): p ∧ q */
  code_block;
  /* post (3): r */
end /* while */
/* post (4): r ∧ ¬q */
```

For this to be valid we need $r \Rightarrow p$: for the first time through the while loop, at precondition (2) we must have both p and q true. Assuming that the block from precondition (2) to postcondition (3) is valid, at the end of the loop r is true. Control is then passed back to the start of the while loop, and condition q is again tested. If q is then true, we re-enter the while loop. At this point we know that $r \wedge q$ must be true, but we want $p \wedge q$ to be true. So we need $r \Rightarrow p$.

At the end of the loop, q must be false. On the other hand, r must be true, so $r \wedge \neg q$.

There is one thing that this does *not* do: prove that the loop actually terminates. Without this we only have a proof of *partial correctness*.

Here is a simple example of how to use these ideas for showing correctness of an algorithm for computing $\max \{a_1, a_2, \ldots, a_n\}$:

```
/* pre: n = length(a) > 0 */
i ← 1
maxval ← a₁
/* pre: maxval = max {a₁, a₂, ..., aᵢ} */
while ( i ≠ n )
  /* pre: maxval = max {a₁, a₂, ..., aᵢ} ∧ i ≠ n */
  if aᵢ > maxval then maxval ← aᵢ
  /* post: maxval = max {a₁, a₂, ..., aᵢ₊₁} */
  i ← i + 1
  /* post: maxval = max {a₁, a₂, ..., aᵢ} */
end while
/* post: maxval = max {a₁, a₂, ..., aᵢ} ∧ i = n */
```

Here we take p to be the statement "$maxval = \max\{a_1, \ldots, a_i\}$", q the statement "$i \neq n$", and r is the same as p.

Recursion is where a function calls itself. Proving things about recursive functions is based on the idea of induction. Consider a recursive function:

```
function my_rec(a, b, ...)
  /* pre: P(a, b, ...) */
```

```
block_of_code_1
/* pre:  P(a',b',...) */
x ← my_rec(a',b',...)
/* post:  R(x,a',b',...) */
block_of_code_2
/* post:  R(y,a,b,c,...) */
return y
end function
```

The precondition that holds at the beginning of the function should also hold just before the recursive call for the inputs for the call. The postcondition that holds at the end of the function should also hold just after the recursive call for the inputs for the call. Then when we call the recursive function, we have the valid code:

```
/* pre:  P(a,b,...) */
x ← my_rec(a,b,...)
/* post:  R(x,a,b,...) */
```

For an example, we can compute Fibonacci numbers recursively via the formula $F_{n+1} = F_n + F_{n-1}$ and the starting values $F_1 = 1$ and $F_2 = 1$. Here is the code:

```
function fib(n)
    /* pre:  n integer */
    if ( n ≤ 0 )
        F ← 0 /* define F_n = 0 for n ≤ 0 */
    else if ( n = 1 )
        F ← 1
    else if ( n = 2 )
        F ← 1
    else /* pre:  n > 2 integer */
        F ← fib(n − 1) + fib(n − 2)
        /* post:  F = F_{n-1} + F_{n-2} */
    end if
    /* post:  F = F_n */
    return F
end function fib
```

It is easy with this example to see that it is correct (although the assignment $F \leftarrow 0$ might be replaced by raising an error for $n \leq 0$).

2.7.1   Correctness by design*

Proofs of correctness can also be used to *design* algorithms. Below is an example of how we can use proof of correctness ideas to design a sorting algorithm. In logical terms, what does a sorting algorithm do? It operates on an array a of length n so that the values after sorting are a permutation of the original values, and $(1 \le i < j \le n) \Rightarrow (a_i \le a_j)$. The most commonly used algorithm for sorting is the quicksort algorithm. The basic idea of the algorithm is to pick a value v from the list, and then to partition the list with index p so that $(i \le p) \Rightarrow (a_i \le v)$ and $(i > p) \Rightarrow (a_i > v)$. Then we can recursively partition the two parts of the array: a_i for $i \le p$ and a_i for $i > p$. For convenience, let us define "sorted(a, i, j)" be the statement "$\forall k(i \le k < j \Rightarrow a_k \le a_{k+1})$." The quicksort algorithm can be put into recursive form like this:

```
function quicksort(a, i, j)
   /* pre: 1 ≤ i ≤ j ≤ length(a) */
   if ( i = j )
      return a
   else /* i < j */
      /* pre 1: 1 ≤ i ≤ j ≤ length(a) */
      b ← copy(a)
      p ← partition(a, i, j)
      /* post 1: i ≤ p < j,
          a is a permutation of b,
          ∀k(¬(i ≤ k ≤ j) ⇒ aₖ = bₖ),  and
         ∀k ∀ℓ((i ≤ k ≤ p) ∧ (p + 1 ≤ ℓ ≤ j) ⇒ aₖ ≤ aₗ)  */
      c ← copy(a)
      quicksort(a, i, p)
      /* post 2: a is a permutation of b,
          ∀k(¬(i ≤ k ≤ p) ⇒ aₖ = cₖ),
          and sorted(a, i, p)  */
      d ← copy(a)
      quicksort(a, p + 1, j)
      /* post 3: a is a permutation of b,
          ∀k(¬(p + 1 ≤ k ≤ j) ⇒ aₖ = dₖ),
          and sorted(a, p + 1, j)  */
   end if
   /* post 4: a is a permutation of b,
       ∀k(¬(i ≤ k ≤ j) ⇒ aₖ = bₖ),
```

```
        and sorted(a, i, j) */
end function quicksort
```

The lines $b \leftarrow a$, $c \leftarrow a$ and $d \leftarrow a$ means that the assertions can refer to the values of the array a at previous points in the code.

We want to show that we really have sorted(a, i, j) at postcondition 4. Since sorted(a, i, p) holds at postcondition 2, sorted$(a, p+1, j)$ holds at postcondition 3, and $a_p \leq a_{p+1}$ from postcondition 1, we have $a_i \leq a_{i+1} \leq \cdots \leq a_{p-1} \leq a_p \leq a_{p+1} \leq a_{p+2} \leq \cdots \leq a_j$, at postcondition 4 as we wanted.

To show that a is a permutation of b, we are using the fact that if a is a permutation of b and c is a permutation of b, then a is a permutation of c. Proving this is basically a matter of unwrapping the definition of "*is a permutation of*".

But there is a missing piece here: we have not shown correctness of *partition*. But we do not even have the code for it. We need to design this routine. The routine needs to satisfy the following assertions

```
function partition(i, j, a)
    /* pre: 1 ≤ i ≤ j ≤ length(a) */
    b ← copy(a)
    . . .
    p ← . . .
    . . .
    /* post: (i ≤ k ≤ p) ∧ (p + 1 ≤ ℓ ≤ j) ⇒ a_k ≤ a_ℓ
          and ∀k(¬(i ≤ k ≤ j) ⇒ a_k = b_k)
          and a is a permutation b */
end function partition
```

Let's start with a piece of the solution: swapping a pair of entries in an array:

```
function swap(a, i, j)
    /* pre: 1 ≤ i, j ≤ length(a) */
    b ← copy(a)
    t ← a_i;   a_i ← a_j;   a_j ← t
    /* post: ∀k(k ≠ i ∧ k ≠ j ⇒ a_k = b_k)
          and a_i = b_j and a_j = b_i */
end function swap
```

Clearly at the postcondition of *swap*, the array a is a permutation of b with only entries i and j swapped.

The *partition* function has to swap entries of the array a so that there is a number p where $i \leq k \leq p$ implies $a_k \leq a_p$ and $p + 1 \leq k \leq j$ implies $a_p \leq a_k$. We do not need to know what p is in advance. In fact, being able to pick p in advance amounts to finding the $(p - i + 1)$th entry of an array. Instead of trying to pick p at the outset, we ought to (perhaps) pick a value $v = a_k$ for some k, and then go through the array a swapping elements according to whether the element is less than or greater than v, tracking where these entries go. We can inductively build up two lists, one from the "bottom" of the array with the smaller elements, and another from the "top" of the array. If we take $\ell = i$ then the bottom list will have indexes $i + 1$, $i + 2$, ..., ℓ and the top list will have indexes r, $r + 1$, ..., j. Initially $\ell = i$ and $r = j + 1$ indicating empty bottom and top lists. When $r = \ell + 1$ the partition is completed, and we can set $p = \ell$.

Set $v = a_i$. We repeatedly increment the left index ℓ until $a_\ell \geq v$, and repeatedly decrease the right index r until $a_r \leq v$. Then we swap a_ℓ and a_r, and repeat the process.

Here is an implementation:

```
function partition(a, i, j)
   /* pre: 1 ≤ i, j ≤ length(a) */
   b ← copy(a);  ℓ ← i;  r ← j + 1;  v ← aᵢ
   while ℓ < r
      /* pre: a is a permutation of b,  ℓ < r,  and
      (i ≤ k ≤ ℓ ⇒ aₖ ≤ v) ∧ (r ≤ k ≤ j ⇒ aₖ ≥ v) */
      ℓ ← ℓ + 1;  while ℓ < r ∧ aₗ < v: ℓ ← ℓ + 1 end
      /* post 1: ∀k(i ≤ k ≤ ℓ − 1 ⇒ aₖ ≤ v)
            and aₗ ≥ v ∨ ℓ = r */
      r ← r − 1;  while ℓ < r ∧ aᵣ > v: r ← r − 1 end
      /* post 2: ∀k(r + 1 ≤ k ≤ j ⇒ aₖ ≥ v)
            and aᵣ ≤ v ∨ ℓ = r */
      swap(aₗ, aᵣ)
      /* post 3: ∀k(i ≤ k ≤ ℓ ⇒ aₖ ≤ v)
            and ∀k(r ≤ k ≤ j ⇒ aₖ ≥ v) */
   end while ℓ < r
   p ← ℓ
   /* post 4: a is a permutation of b,
      and ∀k∀m((i ≤ k ≤ p) ∧ (p + 1 ≤ m ≤ j)
            ⇒ aₖ ≤ v ≤ aₗ) */
   return p
end function partition
```

Postconditions 1 and 2 hold from the rules for `while` loops, together with the fact that ℓ and r are changed by just one in each loop so that "$\ell \geq r$" can be strengthened to "$\ell = r$". Postcondition 3 can be inferred from postconditions 1 and 2 and the swap. Postcondition 4 follows from postcondition 3 and $p = \ell = r$. Full correctness of *partition* can be proven as $r - \ell$ decreases by at least one with each iteration.

2.8 Exercises

(1) In practice, proofs by contradiction often involve several hypotheses: We might want to prove $p \wedge q \wedge r \Rightarrow s$. We only have to show that the negation of the conclusion $\neg s$ implies the negation of *one* of the hypotheses, assuming the other hypotheses are true. Show this formally by demonstrating that $(p \wedge q \wedge r \Rightarrow s)$ is equivalent to $(\neg s \wedge q \wedge r \Rightarrow \neg p)$.

(2) Show that $(p \Rightarrow r) \Rightarrow (((p \Rightarrow (r \Rightarrow s)) \Rightarrow (p \Rightarrow s))$ is a tautology. This can be done directly in terms of the truth table.

(3) Show that the negation of $\exists! x \, P(x)$ is

$$\forall x \, (\neg P(x) \vee \exists y (P(y) \wedge y \neq x)) \, .$$

(4) At the beginning of the chapter there is the syllogism "All men are mortal. Aristotle is a man. Therefore, Aristotle is mortal." Using the predicates $\text{Man}(x)$ representing "x is a man", $\text{Mortal}(x)$ representing "x is mortal", and the constant "Aristotle", write this syllogism as a sequence of three logical statements in predicate calculus. Show that the inference is a valid one.

(5) Derive complete induction from standard induction: For a predicate $P(z)$, let $Q(x)$ be the formula $\forall y(y < x \Rightarrow P(y))$. Show that the standard axiom of induction for Q $([Q(1) \wedge \forall x(Q(x) \Rightarrow Q(x + 1))] \Rightarrow \forall x \, Q(x))$ implies the complete axiom of induction for P: $[P(1) \wedge \forall x \, (\forall y[y < x \Rightarrow P(y)] \Rightarrow P(x))] \Rightarrow \forall x \, P(x)$.

(6) Prove by induction: $3 \times 2^n + 4 \times 9^n$ is divisible by 7 for $n = 0, 1, 2, 3, \ldots$.

(7) Prove by induction: for $n = 2, 3, 4, \ldots$, $n! \geq 2^{n-1}$.

(8) Prove by induction: for $n = 1, 2, 3, \ldots$, $\begin{bmatrix} F_{n+1} & F_n \\ F_n & F_{n-1} \end{bmatrix} = \begin{bmatrix} 1 & 1 \\ 1 & 0 \end{bmatrix}^n$.

(9) Use the results of Exercise 2.8 and induction to show that if m divides n, then F_m divides F_n.

(10) Prove that the set $\mathbb{N} = \{1, 2, 3, 4, \ldots\}$ is a *well-ordered set*; that is, any non-empty set $S \subseteq \mathbb{N}$ has a minimal element $\hat{s}$ where $\hat{s} \leq s$ for all $s \in S$. [**Hint:** Proceed via a proof by contradiction: suppose $S \subset \mathbb{N}$ does not have a minimal element. Then show by induction that $x \notin S$ is true for all $x \in \mathbb{N}$.]

(11) Recall the definitions of *one-to-one* and *onto* from Section 2.5.3. The *composition* of the functions $f \colon A \to B$ and $g \colon B \to C$ is the function $g \circ f \colon A \to C$ where $g \circ f(a) = g(f(a))$.

(a) Show that if f and g are one-to-one, then so is $g \circ f$.
(b) Show that if f and g are onto, then so is $g \circ f$.
(c) Show that if $g \circ f$ is one-to-one, then so is f.
(d) Show that if $g \circ f$ is onto, then so is g.

(12) A set $S \subset \mathbb{R}^n$ is said to be convex if for all $\mathbf{x}$ and $\mathbf{y}$, $\mathbf{x}$, $\mathbf{y} \in S$ and $0 \leq \theta \leq 1$ imply that $\theta \mathbf{x} + (1 - \theta)\mathbf{y} \in S$. Show that if S_1, $S_2 \subset \mathbb{R}^n$ are convex sets, then $S_1 \cap S_2$ is also convex. Give an example of convex sets S_1 and S_2 in $\mathbb{R}^2$ where $S_1 \cup S_2$ is *not* convex. [**Hint:** For the counterexample, first show that a one-point set $S = \{\mathbf{z}\}$ is a convex set.]

(13) ⚠ Consider a predicate on sets $P(S)$ based on a function $f(x, y, \gamma)$ and a fixed set C in the following way: $P(S)$ means

$$\forall x \, \forall y \, \forall \gamma \, [x \in S \wedge y \in S \wedge \gamma \in C \Rightarrow f(x, y, \gamma) \in S].$$

Show that $P(S_1) \wedge P(S_2) \Rightarrow P(S_1 \cap S_2)$. Also show how this implies the first part of Exercise 2.12.

(14) ⚠ For the predicate $P(S)$ from the previous exercise, show that for a family $S_\alpha, \alpha \in J$ of sets satisfying $P(S_\alpha)$ we have $P(\cap_{\alpha \in J} S_\alpha)$. That is, show that $\forall \alpha \, [\alpha \in J \Rightarrow P(S_\alpha)] \Rightarrow P(\cap_{\alpha \in J} S_\alpha)$. We can define the convex hull of a set S to be the intersection of all convex sets containing S: co $S = \cap_C C$ where C ranges over all convex sets where $S \subseteq C$. Show that co S is convex.

(15) Let A be a set. Show that the relation "$\subseteq$" ("subset of") is a partial order on the set of subsets $\mathcal{P}(A)$ of A. Note that "$B \subseteq C$" means "$\forall x(x \in B \Rightarrow x \in C)$".

(16) Show that "divides" is a partial order on the natural numbers $\mathbb{N}$.

(17) We say that y is an *immediate successor* to x in a set S with a partial order $\preceq$ if $x \preceq y$, $x \neq y$, but there is no $z \in S$ where $z \neq x$, y and $x \preceq z \preceq y$. Show that with the usual ordering $\leq$, the set of real numbers $\mathbb{R}$ does not have immediate successors, but the set of integers $\mathbb{Z}$ does.

(18) The following algorithm computes the Ackermann–Peter function for non-negative integers m and n:

```
function Ackermann(m, n)
    if m = 0 then
        return n + 1
    else if n = 0 then
        return Ackermann(m - 1, 1)
```

```
        else
            return Ackermann(m − 1, Ackermann(m, n − 1))
        end
    end function
```

Show by induction that the function call $Ackermann(m, n)$ terminates in finite time. You may need to use proof by induction several times over. Note that the values of this function increase extremely rapidly: $Ackermann(3, 3) = 61$, but $Ackermann(4, 4) = 2^{2^{65536}} - 3$.

(19) Spot the flaw in the following "proof".

"Theorem". *All cows are the same color.*

"Proof". We show that any set of n cows have the same color by induction on n.

Base case: This is clearly true for $n = 1$, since all cows in a set of one cow must have the same color.

Suppose true for $n = k$. Show true for $n = k+1$. Suppose we have a set of $S = \{c_1, c_2, \ldots, c_{k+1}\}$ of $k+1$ cows. Then by the induction hypothesis the cows in each of the sets $\{c_1, c_2, \ldots, c_k\}$ and $\{c_2, c_3, \ldots, c_{k+1}\}$ must have the same color. Therefore the cows in the set $\{c_1, c_2, c_3, \ldots, c_k, c_{k+1}\}$ must also have the same color. "$\square$"

(20) ⚠ Show that any formula ϕ in propositional calculus can be written as $(x_{1,1} \lor x_{1,2} \lor \cdots \lor x_{1,m_1}) \land (x_{2,1} \lor x_{2,2} \lor \cdots \lor x_{2,m_2}) \land \cdots \land (x_{k,1} \lor x_{k,2} \lor \cdots \lor x_{k,m_k})$ where $x_{i,j}$ is either p or $\neg p$ for some propositional variable p. This is called *conjunctive normal form*.

(21) Often mathematicians will use the shorthand $\forall x \in S \colon P(x)$ for $\forall x(x \in S \Rightarrow P(x))$, and $\exists x \in S \colon P(x)$ for $\exists x(x \in S \land P(x))$. Show that the negation of $\forall x \in S \colon P(x)$ is $\exists x \in S \colon \neg P(x)$ and vice-versa. Similarly, we can impose conditions on variables in quantifiers, such as $\forall x > 0 \colon P(x)$ means $\forall x(x > 0 \Rightarrow P(x))$, and $\exists x > 0 \colon P(x)$ means $\exists x(x > 0 \land P(x))$. Show that the negation of $\forall x > 0 \colon P(x)$ is $\exists x > 0 \colon \neg P(x)$.

(22) A way of using inequalities to prove the equality of real numbers is the following result: $\forall \epsilon > 0(|a - b| < \epsilon) \Rightarrow a = b$. Prove this via the contrapositive.

(23) The following statement says that the sequence a_n, $n = 1, 2, 3, \ldots$ is convergent:

$$\exists L \, \forall \epsilon > 0 \, \exists N \, \forall n \, [n \geq N \Rightarrow |a_n - L| < \epsilon].$$

What is the negation of this formula? Use this negation to show that $a_n = (-1)^n$, $n = 1, 2, 3, \ldots$ is *not* convergent.

(24) *Logic of an empty universe.* Suppose nothing at all exists. Show that in such a universe every universally quantified formula $\forall x\, P(x)$ is (vacuously) true, but that every existential statement $\exists x\, P(x)$ is false.

(25) What is the fallacy of the following "proof"?

"Theorem". *My dog is amazing.*

"Proof". My dog is black (observation). Some dogs are black. Therefore, my dog is *some* dog!   "$\square$"

(26) The extended real numbers $\overline{\mathbb{R}} = \mathbb{R} \cup \{\infty\}$. We can extend addition to $\overline{\mathbb{R}}$ by the rules $a + \infty = \infty$, $\infty + a = \infty$ for any real a, and $\infty + \infty = \infty$. Show that the following rules hold for any *extended real* numbers a, b, c:

$$a + b = b + a,$$
$$a + (b + c) = (a + b) + c.$$

But show that subtraction cannot be extended to $\overline{\mathbb{R}}$ by showing that $(\infty - \infty) - \infty \neq \infty - (\infty + \infty)$ via the rule "$a - a = 0$".

(27) Prove that the *partition* function above correctly implements the specification defined by the pre- and post-conditions.

(28) Prove that the following code evaluates the polynomial $p(x) = \sum_{k=0}^{n} a_k x^k$ using Horner's nested multiplication method:

```
function Horner(a, n, x)
    /* pre: n > 0 integer, length(a) = n */
    s ← aₙ;  k ← 0
    while k < n
        /* pre: s = Σⁿℓ₌ₙ₋ₖ aℓxˡ⁻⁽ⁿ⁻ᵏ⁾ */
        s ← sx + aₙ₋ₖ₋₁;  k ← k + 1
    end
    return s
end
```

Chapter 3

Discrete and continuous

It is easy to divide mathematics into "discrete mathematics" and "continuous mathematics": discrete mathematics is about whole numbers and discrete objects, continuous mathematics is about real numbers and approximations. Continuous mathematics is about limits, while discrete mathematics is about counting and algebra. Proofs about limits or continuity usually involve statements like "*For every* $\epsilon > 0$ *there is* ..." In fact, the two are intertwined, and one is often used to help the other. Asymptotic results about discrete objects require continuous mathematics: the number of primes $\leq x$ divided by $x / \ln x$ goes to one as $x \to \infty$. And continuous functions have discrete properties: A continuously differentiable function $f \colon [a, b] \to \mathbb{R}$ where $f(x) = 0$ implies $f'(x) \neq 0$, and $f(a) f(b) < 0$, has an odd number of zeros in $[a, b]$.

In this chapter we investigate a number of topics taken from both continuous and discrete mathematics. Some ideas ("calculate the same thing in two different ways", "swapping the order of summation", and "emphasizing the important") are equally applicable to discrete and continuous mathematics. Some ideas ("prime numbers", "divisibility", and "graphs and networks") are especially appropriate for discrete mathematics, while others ("convergence", and "least upper bound property") are especially appropriate for continuous mathematics.

3.1 Inequalities

Inequalities are essential in any understanding of approximations and real numbers. But there is something tactically different about proving inequalities than proving equalities. Every equality is reversible: "$a = b$" means "$b = a$". But this is not so for inequalities. If, in the middle of a proof of "$a \geq b$" we have managed to show "$a \geq c$" where $b > c$, then we cannot progress in a straight line to the conclusion. We may need to go back, and get something else. We may even need to re-write the first part of the proof. This is not to say that we should be scared of any irreversible step. We will have to take irreversible steps, but we should try to limit them. A classic reference on inequalities is Hardy, Littlewood and Pólya (1952).

3.1.1 *Some classic inequalities*

The Euclidean length of a vector $\mathbf{x} = [x_1, x_2, \ldots, x_n]^T$ is $\|\mathbf{x}\|_2 = \sqrt{x_1^2 + x_2^2 + \cdots + x_n^2} = \sqrt{\mathbf{x} \cdot \mathbf{x}}$ where $\mathbf{u} \cdot \mathbf{v} = \sum_{i=1}^{n} u_i v_i$ is the scalar or dot product of vectors $\mathbf{u}$ and $\mathbf{v}$.

Theorem 3.1 (Cauchy–Schwartz). $|\mathbf{x} \cdot \mathbf{y}| \leq \|\mathbf{x}\|_2 \|\mathbf{y}\|_2$ *for any real vectors* $\mathbf{x}, \mathbf{y} \in \mathbb{R}^n$.

The first proof below starts with an apparently unrelated opening move: $\|\mathbf{x} + s\mathbf{y}\|_2^2 \geq 0$ for any real number s. Why this move? The first retort is "It works! You just need to learn the trick!" A closer inspection reveals a few things: $\|\mathbf{u}\|_2^2 = \mathbf{u} \cdot \mathbf{u}$, and there are many rules that you can use on dot products, such as $\mathbf{w} \cdot (\mathbf{u} + \alpha\mathbf{v}) = \mathbf{w} \cdot \mathbf{u} + \alpha\mathbf{w} \cdot \mathbf{v}$. This makes calculations easier. Also, after expanding $\|\mathbf{x} + s\mathbf{y}\|_2^2 = (\mathbf{x} + s\mathbf{y}) \cdot (\mathbf{x} + s\mathbf{y})$, we have an expression with terms involving $\mathbf{x} \cdot \mathbf{x} = \|\mathbf{x}\|_2^2$, $\mathbf{x} \cdot \mathbf{y}$, and $\mathbf{y} \cdot \mathbf{y} = \|\mathbf{y}\|_2^2$, which are the quantities we want to relate. In the spirit of trying this to see where it goes, we begin.

Proof. First we start from

$$0 \leq \|\mathbf{x} + s\mathbf{y}\|_2^2 = (\mathbf{x} + s\mathbf{y}) \cdot (\mathbf{x} + s\mathbf{y}) = \mathbf{x} \cdot \mathbf{x} + s\mathbf{y} \cdot \mathbf{x} + s\mathbf{x} \cdot \mathbf{y} + s\mathbf{y} \cdot s\mathbf{y}$$
$$= \mathbf{x} \cdot \mathbf{x} + 2s\mathbf{x} \cdot \mathbf{y} + s^2\mathbf{y} \cdot \mathbf{y}.$$

This is true for all s. Also, $\mathbf{x} \cdot \mathbf{x} + 2s\mathbf{x} \cdot \mathbf{y} + s^2\mathbf{y} \cdot \mathbf{y}$ is a quadratic function of s. Either $\mathbf{y} = 0$ or $\mathbf{y} \neq 0$.

> *If $\mathbf{y} = 0$ then our quadratic function degenerates to a constant function, which needs separate treatment.*

If $\mathbf{y} = 0$, then $\mathbf{x} \cdot \mathbf{y} = 0$ and $\|\mathbf{y}\|_2 = 0$, so $\mathbf{x} \cdot \mathbf{y} = 0 \leq 0 = \|\mathbf{x}\|_2 \|\mathbf{y}\|_2$, as we wanted to show.

Now we suppose $\mathbf{y} \neq 0$. Let us look for the minimum over s of $\mathbf{x} \cdot \mathbf{x} + 2s\mathbf{x} \cdot \mathbf{y} + s^2\mathbf{y} \cdot \mathbf{y}$: set

$$\frac{d}{ds}\left(\mathbf{x} \cdot \mathbf{x} + 2s\mathbf{x} \cdot \mathbf{y} + s^2\mathbf{y} \cdot \mathbf{y}\right) = 2\mathbf{x} \cdot \mathbf{y} + 2s\mathbf{y} \cdot \mathbf{y} = 0,$$

which gives the minimizing s (if there is one). Put $s = -\mathbf{x} \cdot \mathbf{y}/\mathbf{y} \cdot \mathbf{y}$.

We do not need to prove that $s = -\mathbf{x} \cdot \mathbf{y}/\mathbf{y} \cdot \mathbf{y}$ minimizes the quadratic for the rest of the argument to hold. Instead of "Let us look for the minimum..." we could continue with "Put $s = -\mathbf{x} \cdot \mathbf{y}/\mathbf{y} \cdot \mathbf{y}$" without explaining where this choice comes from.

Substituting this value of s gives

$$0 \le \mathbf{x} \cdot \mathbf{x} + 2\frac{-\mathbf{x} \cdot \mathbf{y}}{\mathbf{y} \cdot \mathbf{y}}\mathbf{x} \cdot \mathbf{y} + \left(\frac{-\mathbf{x} \cdot \mathbf{y}}{\mathbf{y} \cdot \mathbf{y}}\right)^2 \mathbf{y} \cdot \mathbf{y}$$

$$= \mathbf{x} \cdot \mathbf{x} - 2\frac{(\mathbf{x} \cdot \mathbf{y})^2}{\mathbf{y} \cdot \mathbf{y}} + \frac{(\mathbf{x} \cdot \mathbf{y})^2}{\mathbf{y} \cdot \mathbf{y}} = \|\mathbf{x}\|_2^2 - \frac{(\mathbf{x} \cdot \mathbf{y})^2}{\|\mathbf{y}\|_2^2}.$$

Multiplying by $\|\mathbf{y}\|_2^2 > 0$ gives $0 \le \|\mathbf{x}\|_2^2 \|\mathbf{y}\|_2^2 - (\mathbf{x} \cdot \mathbf{y})^2$. That is, $(\mathbf{x} \cdot \mathbf{y})^2 \le \|\mathbf{x}\|_2^2 \|\mathbf{y}\|_2^2$. Since all quantities are ≥ 0 and the square root function is a monotone function, we can take square roots and keep the inequality:

$$|\mathbf{x} \cdot \mathbf{y}| \le \|\mathbf{x}\|_2 \|\mathbf{y}\|_2 \,,$$

as we wanted. $\qquad\square$

This bound cannot be improved to *"For all $\mathbf{x}$ and $\mathbf{y}$, $|\mathbf{x} \cdot \mathbf{y}| \le \alpha \|\mathbf{x}\|_2 \|\mathbf{y}\|_2$"* with $\alpha < 1$: if we put $\mathbf{y} = t\mathbf{x}$ for any real number t, we get $|\mathbf{x} \cdot \mathbf{y}| = \|\mathbf{x}\|_2 \|\mathbf{y}\|_2$. In fact, this is the only time we get equality as it corresponds to the minimum value of $\|\mathbf{x} + s\mathbf{y}\|_2 = 0$ at $s = -t$.

Here is a second proof: we start from $0 \le (a - b)^2 = a^2 - 2ab + b^2$ so $ab \le \frac{1}{2}(a^2 + b^2)$ with equality if $a = b$. We can apply this to $x_i y_i$ to bound the inner product in terms of $x_i^2 + y_i^2$:

$$\mathbf{x} \cdot \mathbf{y} = \sum_{i=1}^{n} x_i y_i \le \sum_{i=1}^{n} \frac{1}{2}\left(x_i^2 + y_i^2\right) = \frac{1}{2}\sum_{i=1}^{n} x_i^2 + \frac{1}{2}\sum_{i=1}^{n} y_i^2 = \frac{1}{2}\|\mathbf{x}\|_2^2 + \frac{1}{2}\|\mathbf{y}\|_2^2.$$

This is not the bound we are looking for. But this bound does not scale properly: replacing $\mathbf{x}$ by $s\mathbf{x}$ and $\mathbf{y}$ by $s^{-1}\mathbf{y}$ gives $s\mathbf{x} \cdot s^{-1}\mathbf{y} = \mathbf{x} \cdot \mathbf{y}$ but the bound becomes $\frac{1}{2}s^2 \|\mathbf{x}\|_2^2 + \frac{1}{2}s^{-2} \|\mathbf{y}\|_2^2$. Optimizing this bound gives the result we want:

Proof. Either $\mathbf{x} = 0$ or $\mathbf{x} \ne 0$. If $\mathbf{x} = 0$ then $\mathbf{x} \cdot \mathbf{y} = 0 \le 0 = \|\mathbf{x}\|_2 \|\mathbf{y}\|_2$, as we wanted.

Similarly, if $\mathbf{y} = 0$ then $\mathbf{x} \cdot \mathbf{y} = 0 \le 0 = \|\mathbf{x}\|_2 \|\mathbf{y}\|_2$ as we wanted.

Let us suppose that $\mathbf{x} \ne 0$ and $\mathbf{y} \ne 0$. Since $0 \le (a - b)^2 = a^2 - 2ab + b^2$ we have $ab \le \frac{1}{2}\left(a^2 + b^2\right)$ for any real a and b. Then for any $s > 0$ we have

$$\mathbf{x} \cdot \mathbf{y} = s\mathbf{x} \cdot s^{-1}\mathbf{y} = \sum_{i=1}^{n} sx_i \, s^{-1}y_i \le \sum_{i=1}^{n} \frac{1}{2}\left(s^2 x_i^2 + s^{-2}y_i^2\right)$$

$$= \frac{s^2}{2}\sum_{i=1}^{n} x_i^2 + \frac{s^{-2}}{2}\sum_{i=1}^{n} y_i^2 = \frac{1}{2}s^2 \|\mathbf{x}\|_2^2 + \frac{1}{2}s^{-2} \|\mathbf{y}\|_2^2.$$

We now minimize the bound:

$$\frac{d}{ds}\left(\frac{1}{2}s^2\|\mathbf{x}\|_2^2 + \frac{1}{2}s^{-2}\|\mathbf{y}\|_2^2\right) = s\|\mathbf{x}\|_2^2 - s^{-3}\|\mathbf{y}\|_2^2 = 0$$

which gives $s = [\|\mathbf{y}\|_2 / \|\mathbf{x}\|_2]^{1/2} > 0$.

Again, we do not need to prove that this value of s minimizes the function, just that this value of s makes sense.

Substituting this value of s into $\mathbf{x} \cdot \mathbf{y} \le \frac{1}{2}s^2\|\mathbf{x}\|_2^2 + \frac{1}{2}s^{-2}\|\mathbf{y}\|_2^2$ gives

$$\mathbf{x} \cdot \mathbf{y} \le \frac{1}{2}\frac{\|\mathbf{y}\|_2}{\|\mathbf{x}\|_2}\|\mathbf{x}\|_2^2 + \frac{1}{2}\frac{\|\mathbf{x}\|_2}{\|\mathbf{y}\|_2}\|\mathbf{y}\|_2^2 = \|\mathbf{x}\|_2\|\mathbf{y}\|_2.$$

Since this is true for all $\mathbf{x}$ and $\mathbf{y}$, we can replace $\mathbf{y}$ with $-\mathbf{y}$:

$$-\mathbf{x} \cdot \mathbf{y} = \mathbf{x} \cdot (-\mathbf{y}) \le \|\mathbf{x}\|_2\|-\mathbf{y}\|_2 = \|\mathbf{x}\|_2\|\mathbf{y}\|_2,$$

so $|\mathbf{x} \cdot \mathbf{y}| \le \|\mathbf{x}\|_2\|\mathbf{y}\|_2,$

which is the bound we want.   □

3.1.2 *Convex functions*

Convex functions have some very nice properties, which can help to prove many classic inequalities.

Definition 3.1. *A function $f\colon \mathbb{R}^n \to \mathbb{R}$ is* convex *if for any $\mathbf{x}, \mathbf{y} \in \mathbb{R}^n$ and $0 \le \theta \le 1$ we have*

$$f(\theta\mathbf{x} + (1-\theta)\mathbf{y}) \le \theta\, f(\mathbf{x}) + (1-\theta)\, f(\mathbf{y}). \tag{3.1}$$

It can also be shown by induction on m that for f convex if $0 \le \theta_1, \theta_2, \ldots, \theta_m$ where $\sum_{i=1}^m \theta_i = 1$, then

$$f(\theta_1\mathbf{x}_1 + \theta_2\mathbf{x}_2 + \cdots + \theta_m\mathbf{x}_m) \le \theta_1\, f(\mathbf{x}_1) + \theta_2\, f(\mathbf{x}_2) + \cdots + \theta_m\, f(\mathbf{x}_m). \tag{3.2}$$

There are several equivalent definitions for convexity where f is smooth, including the well-known one from calculus: a twice differentiable function $f\colon \mathbb{R} \to \mathbb{R}$ is convex if and only if

$$f''(x) \ge 0 \qquad \text{for all } x. \tag{3.3}$$

A continuously differentiable function $f\colon \mathbb{R} \to \mathbb{R}$ is convex if and only if

$$f(y) \ge f(x) + (y-x)\, f'(x) \qquad \text{for all } x, y. \tag{3.4}$$

(Showing that convexity implies (3.4) is Exercise 3.2. The equivalence of (3.3) is Exercise 3.3.) Adding convex functions and multiplying by positive constants give further convex functions.

The condition "$f(y) \ge f(x) + (y-x)\, f'(x)$ for all x, y" easily leads to:

Theorem 3.2. *For a continuously differentiable convex function $f : \mathbb{R} \to \mathbb{R}$, x^* is a global minimizer if and only if $f'(x^*) = 0$.*

The condition that $f'(x^*) = 0$ at a minimizer x^* is usually only a necessary, not a sufficient condition. But if f is convex, $f'(x^*) = 0$ is both necessary and sufficient.

Proof. If x^* is a minimizer, then $f(y) \geq f(x^*)$ for all y. In particular, if $y = x^* + h$, then $f(x^* + h) \geq f(x^*)$. For $h > 0$, $(f(x^* + h) - f(x^*))/h \geq 0$ and so

$$f'(x^*) = \lim_{h \to 0^+} (f(x^* + h) - f(x^*))/h \geq 0.$$

For $h < 0$, $(f(x^* + h) - f(x^*))/h \leq 0$ and so

$$f'(x^*) = \lim_{h \to 0^-} (f(x^* + h) - f(x^*))/h \leq 0.$$

Thus $0 \geq f'(x^*) \geq 0$ and so $f'(x^*) = 0$, as we wanted.

If $f'(x^*) = 0$, then for all y we have $f(y) \geq f(x^*) + (y - x^*)f'(x^*) = f(x^*)$, so x^* is a global minimizer. $\qquad\square$

We can use this to prove some inequalities.

Theorem 3.3 (Young). *For all a, $b \geq 0$, and p, $q > 1$ with $p^{-1} + q^{-1} = 1$, we have*

$$ab \leq \frac{a^p}{p} + \frac{b^q}{q}.$$

Proof. Let $f(a) = |a|^p /p - ab$.

We aim to show that $f(a) \geq -b^q/q$.

This is a convex function since $f''(a) = (p - 1)a^{p-2} > 0$ for $a > 0$ since $p > 1$. Note that $f'(a) = \text{sgn}(a)\, |a|^{p-1} - b$ with $f'(0) = -b$. Then $f'(a) = 0$ implies that $a > 0$ and that $a^{p-1} = b$; that is, $a = b^{1/(p-1)}$. Thus the minimizer of f is $a^* = b^{1/(p-1)}$, and $f(a) \geq f(a^*) = b^{p/(p-1)}/p - b^{1/(p-1)}b = b^{p/(p-1)}/p - b^{p/(p-1)}$. Note that $p/(p-1) = 1/(1-p^{-1}) = 1/(q^{-1}) = q$ since $p^{-1} + q^{-1} = 1$. Then $f(a^*) = b^q/p - b^q = -b^q(1 - p^{-1}) = -b^q\, q^{-1} = -b^q/q$.

Therefore, $a^p/p - ba = f(a) \geq f(a^*) = -b^q/q$, and so $a^p/p + b^q/q \geq ab$ as we wanted. $\qquad\square$

Another celebrated result is the inequality between the arithmetic and geometric means of a set of numbers:

Theorem 3.4 (Inequality of arithmetic and geometric means). *If a_1, a_2, ..., a_n are positive numbers, then*

$$\frac{1}{n}(a_1 + a_2 + \cdots + a_n) \geq (a_1 a_2 \cdots a_n)^{1/n}.$$

The expression on the left ($n^{-1}(a_1 + a_2 + \cdots + a_n)$) is the arithmetic mean, and the expression on the right ($(a_1 a_2 \cdots a_n)^{1/n}$) is the geometric mean. Proving the inequality is easy once you realize that $f(x) = -\ln x$ is a convex function. Why this function? Because logarithms connect addition with multiplication, and the minus sign makes it convex.

Proof. The function $f(x) = -\ln x$ is a convex function on the infinite interval $(0, \infty)$ since $f''(x) = +x^{-2} > 0$ for $x > 0$. So for a_1, a_2, ..., $a_n > 0$ and $\theta_1 = \theta_2 = \cdots = \theta_n = 1/n$ we can apply (3.2) to get

$$f(\theta_1 a_1 + \theta_2 a_2 + \cdots + \theta_n a_n) \leq \theta_1 f(a_1) + \theta_2 f(a_2) + \cdots + \theta_n f(a_n);$$

that is,

$$-\ln(n^{-1}(a_1 + a_2 + \cdots + a_n)) \leq -n^{-1}(\ln a_1 + \ln a_2 + \cdots + \ln a_n).$$

Reversing signs reverses the inequality:

$$\ln(n^{-1}(a_1 + a_2 + \cdots + a_n)) \geq n^{-1}(\ln a_1 + \ln a_2 + \cdots + \ln a_n).$$

Taking exponentials of both sides (and using the fact that the exponential function is monotone increasing) we get

$$n^{-1}(a_1 + a_2 + \cdots + a_n) \geq \exp\left(n^{-1}(\ln a_1 + \ln a_2 + \cdots + \ln a_n)\right)$$
$$= \exp(\ln a_1 + \ln a_2 + \cdots + \ln a_n)^{1/n}$$
$$= (a_1 a_2 \cdots a_n)^{1/n},$$

as we wanted. $\square$

This not only applies to sums, but also to integrals: if $\theta(x) \geq 0$ for all x and $\int_a^b \theta(x)\,dx = 1$, then for f convex and g continuous,

$$f\left(\int_a^b g(x)\,\theta(x)\,dx\right) \leq \int_a^b f(g(x))\,\theta(x)\,dx,$$

which is known as *Jensen's inequality*.

3.2 Some proofs in number theory

Number theory is an extremely old part of mathematics, and goes back at least to the Babylonians. But even today it is a vital and growing part of mathematics with new ideas, techniques and conjectures. Number theory was the basis for more abstract developments, such as abstract algebra. It is also connected to analysis in a number of ways such as through the Riemann zeta function:

$$\zeta(s) = \sum_{n=1}^{\infty} \frac{1}{n^s} = \prod_{p \text{ prime}} \frac{1}{1 - p^{-s}}$$

which is well-defined for complex s with $\operatorname{Re} s > 1$. The books Hardy, Wright and Wiles (2008) and Pommersheim, Marks and Flapan (2010) are introductions to this fascinating area.

3.2.1 *Division algorithm and gcd*

Divisibility is a basic concept of number theory: for example, 3 divides 12, but 4 does not divide 6.

Definition 3.2. *An integer m divides an integer n (denoted "$m \mid n$") means that there is an integer k such that $n = k\,m$. The greatest common divisor (gcd) of positive integers m and n is d where $d \mid m$ and $d \mid n$ but for all integers d', $(d' \mid m \wedge d' \mid n) \Rightarrow |d'| \leq d$.*

Divisibility is a partial order relation for positive integers: if $a \mid b$ and $b \mid c$ then $a \mid c$; if $a \mid b$ and $b \mid a$ then $a = b$ for positive integers a, b, and c. Also $a \mid a$ for any positive integer a since $a = 1\,a$. Also for positive integers, $a \mid b$ implies $1 \leq a \leq b$.

For general integers, $a \mid b$ and $b \mid a$ implies $a = \pm b$.

Division of positive integers with remainder is something we did in high school, but now let us prove that it can be done in general:

Theorem 3.5 (Division algorithm). *If a, b are non-negative integers with $b > 0$, then there are integers $q \geq 0$ and r with $0 \leq r < b$ where*

$$a = q\,b + r.$$

Furthermore, given a and b, q and r are unique.

The proof of existence of q and r is by (complete) induction. Should it be induction on a or on b? Induction on a is easier since $a' = a - b = (q-1)b + r = q'\,b + r$.

Proof. We first show the *existence* of q and r satisfying $a = q\,b + r$ with $q \geq 0$ and $0 \leq r < b$ by complete induction on a. Note that this proves the result for $a \geq 0$.

Base case: For $a = 0$ we have $0 = 0\,b + 0$ so $q = r = 0$. Then $q \geq 0$ and $0 \leq r < b$.

Suppose true for all $0 \leq c < a$. Show true for a: Either $a < b$ or $a \geq b$.

For the case $a < b$ we set $a = 0\,b + a$ ($q = 0$ and $r = a$) which satisfy $q \geq 0$ and $0 \leq r < b$.

For the case $a \geq b$, then set $c = a - b \geq 0$. By the induction hypothesis, there are $q' \geq 0$ and $0 \leq r' < b$ where $c = q'\,b + r'$. So $a = c + b = (q' + 1)\,b + r'$. Setting $q = q' + 1$ and $r = r'$ gives $a = q\,b + r$ where $q \geq 0$ and $0 \leq r < b$.

Conclusion: For all non-negative integers a there are $q \geq 0$ and $0 \leq r < b$ such that $a = q\,b + r$.

To prove *uniqueness*, suppose $a = q_1 b + r_1 = q_2 b + r_2$ with $q_1, q_2 \geq 0$ and $0 \leq r_1, r_2 < b$. Subtracting gives $0 = (q_1 - q_2)b + (r_1 - r_2)$; that is, $r_2 - r_1 = (q_1 - q_2)b$. This means that $b \mid (r_1 - r_2)$. That is, there is an integer s where $r_1 - r_2 = b\,s$. Taking absolute values gives $|r_1 - r_2| = b\,|s|$. Now $0 \leq |r_1 - r_2| < b$ since $-b < r_1 - r_2 < +b$. If $|r_1 - r_2| \neq 0$ then it is a positive integer which is divisible by b, so $b \leq |r_1 - r_2|$, which is a contradiction. Therefore $|r_1 - r_2| = 0$ and so $r_1 = r_2$. We then have $0 = (q_1 - q_2)b$, and since $b > 0$, we get $q_1 - q_2 = 0$. That is, $r_1 = r_2$ and $q_1 = q_2$, and so there is only one $q \geq 0$ and $0 \leq r < b$ where $a = q\,b + r$. □

There are a few things to note:

- Existence requires induction to prove, while uniqueness just requires inequalities and divisibility.
- The theorem is called the proof of the division *algorithm*. What is the algorithm? We can unwrap the induction proof to turn it into an algorithm we can compute with:

```
function division(a, b)
    /* returns pair (q, r):  a = qb + r,  0 ≤ r < b */
    /* pre:  a ≥ 0 and b > 0 */
    if ( a < b )
        /* post 1:  a = 0b + a,  0 ≤ a < b */
        return (0, a)
    else /* a ≥ b */
        (q', r') ← division(a − b, b)
        /* post 2:  a − b = q'b + r',  0 ≤ r' < b */
```

```
        return (q' + 1, r')
     end
  end
```

Now we will look at the greatest common divisor of a pair of positive integers $d = \gcd(a, b)$. The greatest common divisor is the positive integer d where $d \mid a$ and $d \mid b$, and for any positive integer f where $f \mid a$ and $f \mid b$ we have $d \geq f$. The existence of the greatest common divisor comes from Euclid's algorithm:

```
function gcd(a, b)
   /* pre: a, b > 0 integers */
   (q, r) ← division(a, b)
   /* post: a = qb + r and 0 ≤ r < b */
   if r ≠ 0
      return gcd(b, r)
   else /* r = 0 */
      return b
   end
end
```

Euclid's algorithm must terminate since in the recursive call $gcd(b, r)$ we have $0 \leq r < b$ so the second argument must decrease with each recursive call. But is $gcd(a, b)$ really the greatest common divisor of a and b? First we check that it is indeed a common divisor.

Lemma 3.1. *If $d = gcd(a, b)$ for positive integers a, b, then $d \mid a$ and $d \mid b$.*

Proof. This is by complete induction on b.

Base case: $b = 1$: Then $division(a, b)$ returns $(a, 0)$ since $a = a \times 1 + 0$ and $0 < 1$. Then the returned value is $b = 1$, which indeed divides both a and $b = 1$ for any positive integer a.

Suppose true for $b < m$; show true for $b = m$: Then $division(a, m)$ returns (q, r) where $a = qm + r$ and $0 \leq r < m$. We now have two separate cases to consider: $r = 0$ or $r \neq 0$.

If $r = 0$ then the returned value is m. On the other hand, $r = 0$ so $a = qm$ and $m \mid a$. Since $m \mid m$, it is clear that m divides both a and m.

If $r \neq 0$, let $f = gcd(m, r)$ from the recursive call. By the induction hypothesis, $f \mid m$ and $f \mid r$. As f, m and r are positive integers, there are further positive integers g and h where $m = gf$ and $r = hf$. Then

$a = qm + r = qgf + hf = (qg + h)f$ and $f \mid a$. Thus f is a common divisor of a and m, as we wanted.   □

The next question is, how do we know that $gcd(a, b)$ the *greatest* common divisor? There is a quick test we can use:

Lemma 3.2. *If there are integers (not necessarily positive) s and t where $d = sa + tb$, then any common divisor of a and b is a divisor of d.*

Proof. If f is a common divisor of a and b, then we can write $a = uf$ and $b = vf$ for integers u and v. Then $d = sa + tb = suf + tvf = (su + tv)f$ so f divides d.   □

We can use this Lemma with Euclid's algorithm to show that the returned d is indeed the greatest common divisor.

Theorem 3.6. *If $d = gcd(a, b)$ for positive integers a and b, then d is the greatest common divisor of a and b.*

Proof. We have to prove by complete induction on b that there are integers s and t so that $d = sa + tb$.

Base case: $b = 1$: If $b = 1$ then $q = a$ and $r = 0$, so $d = b = 0a + 1b = sa + tb$ with $s = 0$ and $t = 1$.

Suppose true for $b < m$; show true for $b = m$: Then $division(a, m)$ returns (q, r) where $a = qm + r$ and $0 \leq r < m$. Again we have two separate cases to consider: $r = 0$ and $r \neq 0$.

If $r = 0$ then the returned value is $d = m = sa + tm$ with $s = 0$ and $t = 1$.

If $r \neq 0$, then $d = gcd(m, r)$. By the induction hypothesis, there are integers s' and t' where $d = s'm + t'r = s'm + t'(a - qm) = sa + tm$ where $s = t'$ and $t = s' - qt'$.

Thus in either case, there are integers s and t where $d = sa + tm$ where $d = gcd(a, m)$.

By complete induction, if $d = gcd(a, b)$ then there are integers s and t where $d = sa + tb$.

By Lemma 3.1, $d \mid a$ and $d \mid b$. We have just shown that there are integers s and t where $d = sa + tb$. Then by Lemma 3.2, if f is a positive common divisor of a and b, then f is a divisor of d. So $d = wf$ for some (positive) integer w; as $w \geq 1$ this implies that $d \geq f$. Thus $d \geq f$ whenever f is a common divisor of a and b. That is, d is the *greatest* common divisor of a and b, as we wanted.   □

A consequence of the representation $gcd(a, b) = sa + tb$ is the following result:

Lemma 3.3. *If $a \mid bc$ and $gcd(a, b) = 1$ then $a \mid c$.*

Proof. If $gcd(a, b) = 1$ then there are integers s and t where $sa + tb = 1$. On the other hand, $a \mid bc$ means that there is a k where $bc = ka$. So $tka = tbc = (1 - sa)c = c - sca$. Adding sca to both sides gives $(tk + sc)a = c$ and therefore, $a \mid c$, as we wanted. □

3.2.2 Prime numbers

Definition 3.3. *A positive integer p is a* prime number *if $p \neq 1$ and the only divisors of p are one and p.*

Note that one is excluded from being a prime number.

A fundamental result that most students are aware of is the following theorem, which is known as the *Fundamental Theorem of Arithmetic*:

Theorem 3.7. *For any positive integer $a > 1$ there is a unique factorization*

$$a = p_1^{\alpha_1} p_2^{\alpha_2} \cdots p_m^{\alpha_m}$$

where $p_1 < p_2 < \cdots < p_m$ are prime numbers and $\alpha_1, \alpha_2, \ldots, \alpha_m$ are positive integers.

Proof. We first note an immediate consequence of Lemma 3.3: if p is prime and $p \mid bc$ for positive integers b and c, then $p \mid b$ or $p \mid c$: either $gcd(p, b) = 1$ or $gcd(p, b) = p$ since one and p are the only divisors of p. In the first case Lemma 3.3 implies that $p \mid c$; in the second case $p \mid b$. Thus $p \mid b$ or $p \mid c$.

We need to first show that there is a (finite) factorization for any positive integer $a > 1$. We use complete induction on a to show *existence* of a factorization.

Base case: $a = 2$: We start with $a = 2$ since the hypothesis is that $a > 1$. Note that $a = 2^1$ and 2 is prime.

Suppose true for $a < c$; show true for $a = c$: Either c has a divisor $1 < d < c$ or there is no such divisor.

If there is such a divisor them $c = d(c/d)$. By the induction hypothesis, both d and c/d have a factorization by primes, which can be combined to form a factorization of c, as we wanted.

If there is no such divisor, then c is prime and the factorization is $c = c^1$.

Therefore: for any integer $a \geq 2$ there is a factorization $a = p_1^{\alpha_1} p_2^{\alpha_2} \cdots p_m^{\alpha_m}$.

The next task is to show that this factorization is *unique*. We prove this by complete induction on a.

Base case: $a = 2$: Again, 2 is prime, so the only factorization is $a = 2^1$.

Suppose true for $a < c$; show true for $a = c$:

Suppose that

$$c = p_1^{\alpha_1} p_2^{\alpha_2} \cdots p_m^{\alpha_m} = q_1^{\beta_1} q_2^{\beta_2} \cdots q_n^{\beta_n}$$

where $p_1 < p_2 < \cdots < p_m$ and $q_1 < q_2 < \cdots < q_n$ are primes, while $\alpha_1, \ldots, \alpha_m, \beta_1, \ldots, \beta_n$ are positive integers. So $p_1 \mid c = q_1^{\beta_1} q_2^{\beta_2} \cdots q_n^{\beta_n}$. By repeated application of the result that "$p \mid fg \Rightarrow (p \mid f \lor p \mid g)$" for any prime p, we have $p_1 \mid q_j$ for some j. But since q_j is prime and $p_1 > 1$, $p_1 = q_j$. Similarly, $q_1 = p_i$ for some i. If $j > 1$, then $p_1 = q_j > q_1 = p_i \geq p_1$ which is impossible. So $j = 1$; that is, $p_1 = q_1$.

We can divide c by p_1 to get

$$\frac{c}{p_1} = p_1^{\alpha_1 - 1} p_2^{\alpha_2} \cdots p_m^{\alpha_m} = p_1^{\beta_1 - 1} q_2^{\beta_2} \cdots q_n^{\beta_n}.$$

By the induction hypothesis, c/p_1 has a unique factorization or $c/p_1 = 1$. In the case $c/p_1 = 1$ we have the unique factorization $c = p_1^1$, so we now suppose $c/p_1 > 1$. Then uniqueness of the factorization of c/p_1 implies that $\alpha_1 = \beta_1$, $p_2 = q_2$, $\alpha_2 = \beta_2$, etc. That is, the factorization of c is unique, as we wanted.

Therefore: For any integer $a > 1$, there is a unique factorization as a product of prime powers. $\qquad\square$

There are other systems in which there are primes and prime factorization, but not uniqueness of factorization. What makes the proof work for ordinary integers is that we can order the primes in a way that is consistent with divisibility. For a discussion of these issues, see for example Hardy, Wright and Wiles (2008).

3.2.3 *"There are infinitely many primes"*

There are many different proofs of this result with some being very discrete while others are based on calculus. Some proofs are elegant and self-contained, while others are more sophisticated and lead to further ideas and results.

Theorem 3.8. *There are infinitely many prime numbers.*

We start with a classic proof by contradiction.

Proof. Suppose that there are finitely many primes $p_1, p_2, \ldots, p_N$. Then consider the number $a = p_1 p_2 \cdots p_N + 1$. Now a has a factorization into primes $a = q_1^{\beta_1} \cdots q_n^{\beta_n}$, so $q_1 \mid a$. But $gcd(a, p_i) = 1$ for any i by Euclid's algorithm.

Then q_1 does not divide p_i for any i, for this would make q_1 a common divisor of a and p_i and so $q_1 \leq 1$, which is impossible for a prime. Thus $q_1 \neq p_i$ for any i, and q_1 is a new prime, contradicting our assumption that we have all the primes. □

Here is an alternative proof based on divergence of the series $\sum_{n=1}^{\infty} 1/n$.

Proof. If $p > 1$ then

$$\frac{1}{1 - p^{-1}} = \sum_{j=0}^{\infty} p^{-j}.$$

If the set of all primes was $\{p_1, p_2, \ldots, p_N\}$, then since every integer $n > 1$ has a unique prime factorization,

$$\sum_{n=1}^{\infty} \frac{1}{n} = \prod_{p \text{ prime}} \sum_{j=0}^{\infty} p^{-j} = \prod_{k=1}^{N} \frac{1}{1 - p_k^{-1}}.$$

However, $\sum_{n=1}^{\infty} 1/n$ diverges to infinity, and so cannot equal the finite value on the right; this is a contradiction. □

This second proof has some hidden depths to it: the related series $\sum_{n=1}^{\infty} 1/n^s$ does converge for $s > 1$, and we can write

$$\sum_{n=1}^{\infty} \frac{1}{n^s} = \prod_{p \text{ prime}} \sum_{j=0}^{\infty} p^{-js} = \prod_{k=1}^{\infty} \frac{1}{1 - p_k^{-s}} \qquad \text{for } s > 1, \qquad (3.5)$$

where p_k is the kth prime number. However, the left hand side goes to $+\infty$ as s goes down to one. This gives us information about how quickly $p_k \to \infty$ as $k \to \infty$, and therefore how many primes are in a given range.

3.3   Calculate the same thing in two different ways

To show that two things are equal, it is enough to show that they are equal to the same thing. Counting or calculating the same quantity in two different ways will show that two resulting expressions are equal. The main challenge is deciding *what* to calculate in two different ways. Often we can get a good idea about this by looking for something that is related to both expressions in some way. In what follows, if A is a set then $|A|$ is the number of elements of the set A.

3.3.1   *Sums of degrees in graphs*

An undirected graph $G = (V, E)$ consists of a vertex set V, and an edge set E; each edge $e \in E$ is a set of distinct vertices $e = \{u, v\} \subset V$. There is no distinction between $\{u, v\}$ and $\{v, u\}$. More definitions of concepts related to graphs can be found in Definition 1.3 on p. 15.

Definition 3.4. *The* degree *of a vertex $v \in V$ in an undirected graph $G = (V, E)$ is the number of edges containing v. This is denoted by $\deg_G(v)$; if G is understood from context, then we often write $\deg(v)$.*

Theorem 3.9. *If $G = (V, E)$ is a finite undirected graph, then*

$$\sum_{v \in V} \deg_G(v) = 2\,|E|.$$

That is, the sum of the degrees of the nodes of a graph equals twice the number of edges.

If we can find a set where the counting the number of elements gives the quantity on the left, and counting the number of elements in a different way gives the quantity on the right, then the two numbers must be equal. But what set? Clearly it has to relate vertices and edges. One way is to look for suitable sets of pairs (v, e) of vertices v and edges e. It should not be the set of *all* possible pairs (v, e); that would just be the Cartesian product set $V \times E$ which has $|V \times E| = |V|\,|E|$ elements, not $2\,|E|$ elements. Instead, we need to look at a set that clearly involves the structure of the graph. The set that works for us is the set of pairs (v, e) where vertex v belongs to edge e: $\{\,(v, e) \mid v \in e \in E\,\}$.

Proof. Consider the set $S = \{\,(v, e) \mid v \in e \in E\,\}$, the set of pairs of vertices v and edges e where v is contained in (or incident to) e. We count the elements of S in two ways: one by counting how many pairs (v, e) are in S for a given vertex v, the other by counting how many pairs (v, e) are in S for a given edge e.

The first approach gives

$$|S| = \sum_{v \in V} |\{\,e \in E \mid (v, e) \in S\,\}|$$

$$= \sum_{v \in V} |\{\,e \in E \mid v \in e\,\}|$$

$$= \sum_{v \in V} \deg_G(v) \qquad \text{(by definition of } \deg_G(v)\text{)}.$$

The second approach gives

$$|S| = \sum_{e \in E} |\{\, v \in V \mid (v,e) \in S \,\}|$$

$$= \sum_{e \in E} |\{\, v \in V \mid v \in e \,\}|$$

$$= \sum_{e \in E} 2 \qquad \text{(since each edge has exactly two vertices)}$$

$$= 2\,|E|\,.$$

Equating the two final expressions gives the desired result. $\qquad\qquad$ $\square$

A similar example can be found in Exercise 3.10.

3.3.2 *The integral $\int_{-\infty}^{+\infty} \exp(-x^2)\, dx$*

The integral $\int_{-\infty}^{+\infty} \exp(-x^2)\, dx$ cannot be computed symbolically using the Fundamental Theorem of Calculus since there is no formula for the indefinite integral of $\exp(-x^2)$ in terms of the usual elementary functions sin, cos, exp, ln, the usual arithmetic operations $(+,\ -,\ \times,\ /)$ and composition of functions. However, the definite integral can be computed by computing $\int_{-\infty}^{+\infty} \int_{-\infty}^{+\infty} \exp(-x^2 - y^2)\, dx\, dy$ both as a repeated integral, and via polar coordinates:

Theorem 3.10.

$$\int_{-\infty}^{+\infty} \exp(-x^2)\, dx = \sqrt{\pi}.$$

Proof. We compute $\int_{-\infty}^{+\infty} \int_{-\infty}^{+\infty} \exp(-x^2 - y^2)\, dx\, dy$ in two different ways. First, we separate the integrals:

$$\int_{-\infty}^{+\infty} \int_{-\infty}^{+\infty} \exp(-x^2 - y^2)\, dx\, dy$$

$$= \int_{-\infty}^{+\infty} \left(\int_{-\infty}^{+\infty} \exp(-x^2) \exp(-y^2)\, dx \right) dy$$

$$= \int_{-\infty}^{+\infty} \exp(-y^2) \left(\int_{-\infty}^{+\infty} \exp(-x^2)\, dx \right) dy$$

$$= \int_{-\infty}^{+\infty} \exp(-y^2)\, dy \int_{-\infty}^{+\infty} \exp(-x^2)\, dx$$

$$= \left(\int_{-\infty}^{+\infty} \exp(-x^2)\, dx \right)^2.$$

On the other hand, we can use polar coordinates: setting $x = r \cos \theta$ and $y = r \sin \theta$, we have "$dx\, dy = r\, dr\, d\theta$", and

$$
\int_{-\infty}^{+\infty} \int_{-\infty}^{+\infty} \exp(-x^2 - y^2)\, dx\, dy = \int_0^{2\pi} \int_0^\infty \exp(-r^2)\, r\, dr\, d\theta
$$

$$
= 2\pi \int_0^\infty \exp(-r^2)\, r\, dr
$$

$$
= 2\pi \int_0^\infty \exp(-u)\, \frac{1}{2}\, du \qquad (u = r^2)
$$

$$
= 2\pi \frac{1}{2} \left(- \exp(-u) \right)|_{u=0}^{u=\infty}
$$

$$
= \pi (\exp(0) - \exp(-\infty)) = \pi.
$$

Thus

$$
\left(\int_{-\infty}^{+\infty} \exp(-x^2)\, dx \right)^2 = \pi.
$$

Since $\int_{-\infty}^{+\infty} \exp(-x^2)\, dx > 0$ as the integrand is positive, the only solution is

$$
\int_{-\infty}^{+\infty} \exp(-x^2)\, dx = \sqrt{\pi},
$$

as we wanted.   □

In this proof we used shortcut that $\exp(-\infty) = 0$. Of course what we meant is that $\lim_{r \to -\infty} \exp(r) = 0$.

3.4 Abstraction and algebra

Much of mathematics works by *abstraction*. That is, we create a set of rules (or axioms) to be followed by a wide range of general structures. We can often prove things about these general structures, and then apply them to specific situations. But when we apply an abstract result, we need to check that the abstract assumptions are satisfied by the specific example.

For example, a *group* is a set G with a binary operation $*: G \times G \to G$ with the following properties:

- the operation is associative: $(a * b) * c = a * (b * c)$ for all a, b, $c \in G$;
- there is an identity element $e \in G$ where $e * a = a * e = a$ for all $a \in G$;
- every element $a \in G$ has an inverse $a^{-1} \in G$ where $a * a^{-1} = a^{-1} * a = e$.

If, in addition, we have $a * b = b * a$ for all a, $b \in G$ then we say that the operation and the group are *commutative*. (Ordinary multiplication is commutative, but matrix multiplication in general is not.)

Note that to specify a group we not only need to specify what the set G is, but also what the operation "$*$" is.

A particular example of a group is the set of integers under addition; another example is the set of integers modulo n under addition where n is a positive integer. This is not the group we need for the following theorem, but a very closely related one. Note that $a \equiv b \pmod{n}$ means that n divides $b - a$. Note that if $a \equiv c \pmod{n}$ and $c \equiv d \pmod{n}$ then both $a + b \equiv c + d \pmod{n}$ and $ab \equiv cd \pmod{n}$. The following theorem is known as Euler's theorem.

Theorem 3.11. *For a positive integer n, let $\phi(n)$ be the number of elements of the set*

$$\{ a \mid 1 \leq a < n \text{ where } gcd(a, n) = 1 \} .$$

Then if $gcd(a, n) = 1$ we have

$$a^{\phi(n)} \equiv 1 \pmod{n}.$$

This can be proved via more abstract theorems about finite groups below:

Theorem 3.12. *Suppose H is a subgroup of a group G (that is, $H \subseteq G$ and H is a group with the same group operation as G) and both are finite sets. Then the number of elements of H divides the number of elements of G.*

We use the notation $|A|$ to denote the number of elements of a set A. A *partition* P of a set A is a collection of non-empty subsets of A where $\bigcup_{B \in P} B = A$ (every element of A is in some subset B from the partition), and $B_1, B_2 \in P \Rightarrow (B_1 = B_2 \vee B_1 \cap B_2 = \emptyset)$ (any two subsets in the partition are either the same or disjoint). Because all the distinct subsets in the partition are disjoint, $|A| = \sum_{B \in P} |B|$.

Proof. For $g \in G$, let $gH = \{ g * h \mid h \in H \}$. This is called a *coset* of H. We want to show that the collection of sets $\{ gH \mid g \in G \}$ is a partition of G; that is, any two sets from this collection are either the same ($g_1 H = g_2 H$) or disjoint ($g_1 H \cap g_2 H = \emptyset$), and that every $x \in G$ is in a set gH for some $g \in G$.

We do this by contradiction: to show $(g_1 H = g_2 H) \vee (g_1 H \cap g_2 H = \emptyset)$ we show that $(g_1 H \cap g_2 H \neq \emptyset) \Rightarrow (g_1 H = g_2 H)$.

For g_1, $g_2 \in G$, suppose $x \in g_1 H \cap g_2 H \neq \emptyset$. Then there are h_1, $h_2 \in H$ where $x = g_1 * h_1 = g_2 * h_2$. Then using inverses of g_2 and h_1 we have $g_2^{-1} * g_1 =$

$h_2 * h_1^{-1} \in H$ since H is a subgroup. Suppose that $y \in g_2 H$; we want to show that $y \in g_1 H$. If $y = g_2 * k$ with $k \in H$, then

$$y = g_2 * (g_2^{-1} * g_1) * (h_2 * h_1^{-1})^{-1} * k$$
$$= g_1 * (h_1 * h_2^{-1} * k) \in g_1 H$$

since $h_1 * h_2^{-1} * k \in H$. So $g_2 H \subseteq g_1 H$. Swapping the roles of g_1 and g_2 and repeating the argument gives $g_1 H \subseteq g_2 H$. Thus $g_1 H = g_2 H$, as we wanted, and $\{\, gH \mid g \in G \,\}$ is a partition.

To show $\{\, gH \mid g \in G \,\}$ partitions G, note that for any $x \in G$ we have $x \in xH$ since the identity element $e \in H$ and $x = x * e$.

> *Now we want to show that each set in the partition has the same number of elements.*

We show that $|gH| = |H|$. There is the function $H \to gH$ given by $h \mapsto g * h$. This is onto by the definition $gH = \{\, g * h \mid h \in H \,\}$. It is one-to-one since $g * h_1 = g * h_2$ implies $h_1 = g^{-1} * g * h_1 = g^{-1} * g * h_2 = h_2$. Thus H and gH have the same number of elements.

Since $\{\, gH \mid g \in G \,\}$ is a partition of G, $|G|$ is the sum of the of the number of elements of the sets in $\{\, gH \mid g \in G \,\}$.

> *Note that we are* not *saying that $|G| = \sum_{g \in G} |gH|$, which is usually* not *true. There are many $g_1, g_2 \in G$ for which $g_1 H = g_2 H$, but count as only one element in $\{\, gH \mid g \in G \,\}$.*

Since all sets in $\{\, gH \mid g \in G \,\}$ have the same number of elements as H, we have $|G| = |\{\, gH \mid g \in G \,\}| \, |H|$, so $|H|$ divides $|G|$. □

We can go a little further towards Euler's theorem by mentioning the *order* of an element $g \in G$ (denoted order(g)): this is the smallest positive integer m where

$$g^m = \underbrace{g * g * \cdots * g}_{m \text{ times}} = e.$$

Note that $g^0 = e$, and that $g^{-k} = (g^{-1})^k$ so $g^{r+s} = g^r * g^s$ for any integers r and s.

Theorem 3.13. *If G is a finite group, then for any $g \in G$, order(g) divides $|G|$.*

Proof. First, for any $g \in G$, order(g) is a well-defined integer. To see why, note that $\{\, g^k \mid k = 1, 2, 3, \ldots \,\} \subseteq G$ and so is a finite set. Thus there must be positive integers $k < \ell$ where $g^k = g^\ell$. Premultiplying by $(g^{-1})^k$ we get $e = g^{\ell - k}$ so there is a positive integer m where $g^m = e$. Clearly then, there must be a unique smallest such positive integer, and so order(g) is a well-defined positive integer.

Now we need to find a suitable subgroup.

Let $m = $ order(g) and $H = \{ g^k \mid k = 0, 1, 2, \ldots, m - 1 \}$. We now show that H is a group with "$*$" as its binary operation. Now $g^0 = e \in H$ so the identity element is in H. If $0 \le k, \ell < m$, then $g^k * g^\ell = g^{k+\ell}$. If $k + \ell < m$ then clearly $g^k * g^\ell \in H$. If $k + \ell \ge m$, then $0 \le k + \ell - m < m$ (as $k + \ell < 2m$), so $g^k * g^\ell = g^{k+\ell} * e = g^{k+\ell} * g^{-m} = g^{k+\ell-m}$. Thus H is closed under the binary operation "$*$". The inverse of g^k is g^{m-k} since $g^k * g^{m-k} = g^m = e$. So H is a subgroup of G.

From Theorem 3.12, $|H| = m$ divides $|G|$. That is, order(g) divides $|G|$. □

An immediate consequence of this theorem is that if $g \in G$, then $g^{|G|} = (g^{\text{order}(g)})^{|G|/\text{order}(g)} = e^{|G|/\text{order}(g)} = e$.

To prove Theorem 3.11, we can use the abstract Theorem 3.13, but to do so we need to identify a group G with the appropriate binary operation "$*$". Note that $a^{\phi(n)}$ involves multiplication rather than division.

Proof of Euler's theorem, Theorem 3.11. Let G be the set

$$\{ a \in \mathbb{Z} \mid 1 \le a < n \text{ and } \gcd(a, n) = 1 \}$$

with $a * b$ being the remainder of ab after dividing by n. Since both a and b have no common factor with n, neither does ab, and subtracting a multiple of n from ab does not change this.

It is not hard to check that this is an associative binary operation: $*: G \times G \to G$ as if $\gcd(a, n) = 1 = \gcd(b, n)$ then $\gcd(ab, n) = 1$ (any prime dividing ab must divide either a or b). Also, $(a * b) * c = (a * b)c - rn = (ab - sn)c - rn = abc - n(sc+r) \equiv abc \pmod{n}$, and $a*(b*c) = a(b*c) - un = a(bc - vn) - un = abc - n(av + u) \equiv abc \pmod{n}$. But there is only one number between zero and $n - 1$ that is congruent to abc modulo n, so $(a * b) * c = a * (b * c)$; that is, the operation is associative. Also 1 is the identity: $1 * a = a * 1 = a$. If $\gcd(a, n) = 1$ then there is a multiplicative inverse modulo n for a: $xa \equiv 1 \pmod{n}$, so $x * a = 1$ in G. Thus G is a group.

Theorem 3.13 then implies that for any $a \in G$, order(a) divides $|G| = \phi(n)$. Thus

$$\underbrace{a\, a\, a \cdots a}_{\phi(n) \text{ times}} \equiv \underbrace{a * a * a * \cdots * a}_{\phi(n) \text{ times}} = 1 \pmod{n}.$$

That is, $a^{\phi(n)} \equiv 1 \pmod{n}$, as we wanted. □

3.5   Swapping sums, swapping integrals

A common technique in proofs is to swap the order of nested sums or integrals. This can be seen as a version of computing the same thing in two different ways.

3.5.1   *The exponential function*

Most books on calculus first define fractional powers of positive numbers, and then define $e \approx 2.71828\ldots$ by means of a round-about approach of defining the exponential function by $(d/dx)(e^x) = e^x$. But if we start with Taylor series, we can define

$$e^x = \sum_{n=0}^{\infty} \frac{x^n}{n!}. \tag{3.6}$$

If we start this way, then we can quickly show that $e^0 = 1$, and check that e^x is a continuous and differentiable function of x, and that $(d/dx)e^x = e^x$. But it is not clear that e^x satisfies the basic property of any exponential function: $e^{x+y} = e^x\, e^y$. This can be done directly from (3.6). The proof is an exercise in manipulating sums, and using the binomial theorem. The questions of convergence for infinite sums we will avoid here, since the series (3.6) converges absolutely for any x.

Theorem 3.14. *For any real x and y, $e^{x+y} = e^x\, e^y$.*

Proof. We expand e^{x+y} and show that it is equal to $e^x\, e^y$. We assume that the rearrangements will not affect the result since all series involved are absolutely convergent. (This is Exercise 3.30.)

From (3.6),

$$e^{x+y} = \sum_{n=1}^{\infty} \frac{(x+y)^n}{n!}$$

$$= \sum_{n=0}^{\infty} \frac{1}{n!} \sum_{k=0}^{n} \binom{n}{k} x^k\, y^{n-k}$$

$$= \sum_{n=0}^{\infty} \sum_{k=0}^{n} \frac{1}{n!} \frac{n!}{k!\,(n-k)!} x^k y^{n-k}.$$

Now we need to swap the order of the summations. Making the sum over k the outer sum means the inner sum over n has to be for $n = k, k+1, k+2, \ldots$.

Now $\sum_{n=0}^{\infty}\sum_{k=0}^{n}$ is the same as $\sum_{k=0}^{\infty}\sum_{n=k}^{\infty}$ (see Figure 3.1). So

$$
\begin{aligned}
e^{x+y} &= \sum_{k=0}^{\infty}\sum_{n=k}^{\infty}\frac{1}{k!\,(n-k)!}x^k y^{n-k} \\
&= \sum_{k=0}^{\infty}\sum_{\ell=0}^{\infty}\frac{1}{k!\,\ell!}x^k y^{\ell} \qquad (\text{put } \ell = n-k) \\
&= \sum_{k=0}^{\infty}\left(\frac{x^k}{k!}\sum_{\ell=0}^{\infty}\frac{y^{\ell}}{\ell!}\right) = \left(\sum_{k=0}^{\infty}\frac{x^k}{k!}\right)\left(\sum_{\ell=0}^{\infty}\frac{y^{\ell}}{\ell!}\right) \\
&= e^x\,e^y,
\end{aligned}
$$

as we wanted. Note that we can bring the inner sum $\sum_{\ell=0}^{\infty}y^{\ell}/\ell!$ outside the outer sum in the third line above, since the inner sum does not depend on k: the inner sum is therefore a constant multiplier for the sum over k, and can be taken outside the sum over k. □

Swapping summations with fixed limits is easy: $\sum_{i=a}^{b}\sum_{j=c}^{d} = \sum_{j=c}^{d}\sum_{i=a}^{b}$. But if the inner sum has limits that depend on the outer sum's dummy variable, then we have to be more careful. A useful way to think about how to swap summations is to draw a picture like Figure 3.1. In the above proof, the inner sum before the swap was $\sum_{k=0}^{n}$ with the limits $k=0$ and $k=n$. Swapping k and n makes n the variable for the inner sum, and its limits are $n=k$ and "$n=\infty$".

The same technique works for swapping integrals as well.

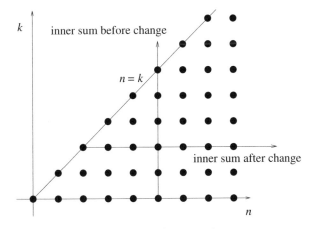

Fig. 3.1: Diagram for swapping sums

3.5.2 *Taylor's theorem with remainder*

A way to get to Taylor's theorem with remainder is to repeatedly apply the fundamental theorem of calculus and to swap the order of double integrals. We can start with

$$f(x) = f(a) + \int_a^x f'(s)\, ds.$$

Now apply the fundamental theorem of calculus to $f'(s)$:

$$f(x) = f(a) + \int_a^x \left[f'(a) + \int_a^s f''(r)\, dr \right] ds$$

$$= f(a) + \int_a^x f'(a)\, ds + \int_a^x \int_a^s f''(r)\, dr\, ds$$

$$= f(a) + f'(a)\,(x - a) + \int_a^x \int_a^s f''(r)\, dr\, ds.$$

To proceed, we swap the order of the double integral $\int_a^x \int_a^s f''(r)\, dr\, ds$. This gives the second order remainder term in integral form:

$$f(x) = f(a) + f'(a)\,(x - a) + \int_a^x (x - r)\, f''(r)\, dr.$$

But we can go further by applying the fundamental theorem of calculus to $f''(r)$, etc.

Here is the full version of Taylor's theorem with remainder in integral form using induction and double integrals.

Theorem 3.15 (Taylor's theorem with integral form remainder). *Provided f is $n + 1$ times continuously differentiable in an interval containing a and x,*

$$f(x) = \sum_{k=0}^{n} \frac{f^{(k)}(a)}{k!}(x - a)^k + \int_a^x \frac{(x - t)^n}{n!} f^{(n+1)}(t)\, dt,$$

where $f^{(k)}$ is the kth derivative of f.

Proof. This is proven by induction on n.

   Base case: $n = 0$: From the fundamental theorem of calculus, $f(x) = f(a) + \int_a^x f'(t)\, dt$ since $(x - t)^0 = 1$ and $0! = 1$.

   Suppose true for $n = m$. Show true for $n = m + 1$. Suppose that

$$f(x) = \sum_{k=0}^{m} \frac{f^{(k)}(a)}{k!}(x - a)^k + \int_a^x \frac{(x - t)^m}{n!} f^{(m+1)}(t)\, dt.$$

Now, from the fundamental theorem of calculus,

$$f^{(m+1)}(t) = f^{(m+1)}(a) + \int_a^t f^{(m+2)}(s)\, ds.$$

So

$$\int_a^x \frac{(x-t)^m}{m!} f^{(m+1)}(t)\,dt$$

$$= \int_a^x \frac{(x-t)^m}{m!} \left[f^{(m+1)}(a) + \int_a^t f^{(m+2)}(s)\,ds \right] dt$$

$$= \int_a^x \frac{(x-t)^m}{m!} f^{(m+1)}(a)\,dt + \int_a^x \frac{(x-t)^m}{m!} \int_a^t f^{(m+2)}(s)\,ds\,dt$$

$$= \frac{f^{(m+1)}(a)}{m!} \int_{x-a}^0 u^m\,(-du) \qquad \text{(put } u = x - t)$$

$$+ \int_a^x \int_a^t \frac{(x-t)^m}{m!} f^{(m+2)}(s)\,ds\,dt$$

$$= \frac{f^{(m+1)}(a)}{m!} \frac{(x-a)^{m+1}}{m+1} + \int_a^x \int_a^t \frac{(x-t)^m}{m!} f^{(m+2)}(s)\,ds\,dt.$$

Now we swap the order of the double integral.

But,

$$\int_a^x \int_a^t \frac{(x-t)^m}{m!} f^{(m+2)}(s)\,ds\,dt$$

$$= \int_a^x \int_s^x \frac{(x-t)^m}{m!} f^{(m+2)}(s)\,dt\,ds$$

$$= \int_a^x f^{(m+2)}(s) \int_s^x \frac{(x-t)^m}{m!}\,dt\,ds$$

$$\text{(since } f^{(m+2)}(s) \text{ does not depend on } t)$$

$$= \int_a^x f^{(m+2)}(s) \int_{x-s}^0 \frac{u^m}{m!}\,(-du)\,ds \qquad \text{(put } u = x - t)$$

$$= \int_a^x f^{(m+2)}(s) \frac{(x-s)^{m+1}}{(m+1)!}\,ds.$$

Therefore,

$$\int_a^x \frac{(x-t)^m}{m!} f^{(m+1)}(t)\,dt = \frac{f^{(m+1)}(a)}{(m+1)!}(x-a)^{m+1}$$

$$+ \int_a^x f^{(m+2)}(s) \frac{(x-s)^{m+1}}{(m+1)!}\,ds.$$

Substituting this into the Taylor's series with remainder of order m gives

$$f(x) = \sum_{k=0}^{m} \frac{f^{(k)}(a)}{k!}(x-a)^k + \frac{f^{(m+1)}(a)}{(m+1)!}(x-a)^{m+1}$$

$$+ \int_a^x \frac{(x-s)^{m+1}}{(m+2)!} f^{(m+2)}(s)\, ds$$

$$= \sum_{k=0}^{m+1} \frac{f^{(k)}(a)}{k!}(x-a)^k + \int_a^x \frac{(x-t)^{m+1}}{(m+2)!} f^{(m+2)}(t)\, dt,$$

as we wanted. $\square$

A variation of this result is Taylor series with a remainder that is simpler to interpret: by the generalized mean value theorem (see Exercise 3.15), if $f^{(m+1)}$ is continuous,

$$f(x) = \sum_{k=0}^{m} \frac{f^{(k)}(a)}{k!}(x-a)^k + f^{(m+1)}(c) \int_a^x \frac{(x-t)^m}{n!}\, dt$$

$$= \sum_{k=0}^{m} \frac{f^{(k)}(a)}{k!}(x-a)^k + f^{(m+1)}(c)\frac{(x-a)^{m+1}}{(m+1)!}$$

for some c between a and x.

3.6   Emphasizing the important

Much of mathematics involves separating out what is important from what is not. This can take various forms: One is to create equivalence classes of items where any pair of items in an equivalence class are the "same" with respect to the important aspects. For example, if we are interested in the colors of a set of balls, we could create equivalence classes containing all balls of the same color. Another way of separating the important items is to separate what is "large" from what is "small"; usually you need to perform exact computations on the "large" part, and bound the "small" part.

3.6.1   *Equivalence relations and well-defined definitions*

An equivalence relation R on a set A has the properties of being transitive ($xRy \wedge yRz \Rightarrow xRz$ for all x, y, and z), symmetric ($xRy \Rightarrow yRx$ for all x and y), and reflexive (xRx for all x). These can be used to focus on a particular characteristic; for example, if you are interested just in color, then we can use the relation $xRy \iff \text{color}(x) = \text{color}(y)$. Equivalence classes are sets of items where any pair of items x, y is in a common equivalence class if and only if xRy. That is, the equivalence classes are the sets

$$[x]_R = \{ y \mid xRy \}.$$

This groups equivalent items into a common sets. The collection of equivalence classes for a set A is $\{ [x]_R \mid x \in A \}$ forms a partition of A: a collection of subsets S of A is a partition if for every S_1, $S_2 \in S$ either $S_1 = S_2$ or $S_1 \cap S_2 = \emptyset$, and the union of all the sets in S is A.

3.6.1.1 *Growth of functions*

We may be interested in the rate of growth of functions. We can use asymptotic notation:

$$f(x) = \mathcal{O}(g(x)) \qquad \text{as } x \to \infty \tag{3.7}$$

means that there are constants L and $C \geq 0$ where

$$x \geq L \quad \Rightarrow \quad |f(x)| \leq C\, g(x). \tag{3.8}$$

This relationship "$f = \mathcal{O}(g)$" is not an equivalence relation; it is reflexive and transitive but not symmetric. We can replace this relation by "$f = \mathcal{O}(g) \wedge g = \mathcal{O}(f)$". This relation is an equivalence relation, and is denoted by "$f = \Theta(g)$". We can use this idea for studying convergence of integrals, for example:

Theorem 3.16. *If f, g are non-negative continuous functions and $f = \Theta(g)$, then the improper integral $\int_0^\infty f(x)\,dx$ is finite if and only if $\int_0^\infty g(x)\,dx$ is finite.*

Proof. Now $f = \Theta(g)$ means $f = \mathcal{O}(g)$ and $g = \mathcal{O}(f)$, so there are constants L_1, L_2 and C_1, $C_2 \geq 0$ where

$$x \leq L_1 \Rightarrow |f(x)| = f(x) \leq C_1\, g(x), \qquad \text{and}$$
$$x \leq L_2 \Rightarrow |g(x)| = g(x) \leq C_2\, f(x).$$

If we put $L = \max(L_1, L_2, 0)$ and $C = \max(C_1, C_2)$, then for $x \geq L$ we have $f(x) \leq C\, g(x)$ and $g(x) \leq C\, f(x)$. So

$$\int_0^\infty f(x)\,dx \leq \int_0^L f(x)\,dx + \int_L^\infty f(x)\,dx$$
$$\leq \int_0^L f(x)\,dx + \int_L^\infty C\, g(x)\,dx$$
$$= \int_0^L f(x)\,dx + C \int_L^\infty g(x)\,dx$$
$$= \int_0^L f(x)\,dx + C \left[\int_0^\infty g(x)\,dx - \int_0^L g(x)\,dx \right]$$

which is finite if $\int_0^\infty g(x)\,dx$ is finite.

On the other hand,

$$\int_0^\infty g(x)\,dx \le \int_0^L g(x)\,dx + C\left[\int_0^\infty f(x)\,dx - \int_0^L f(x)\,dx\right]$$

which is finite if $\int_0^\infty f(x)\,dx$ is finite. $\square$

3.6.1.2 *Modular arithmetic*

Modular arithmetic is another example of how we can use equivalence relations and equivalence classes. We first define "$a \equiv b \pmod{n}$" to mean "n divides $b - a$". This is an equivalence relation. We can then define the equivalence classes

$$[a]_{\equiv n} = \{\, b \mid a \equiv b \pmod{n} \,\}. \tag{3.9}$$

We can perform arithmetic on equivalence classes because we can consistently perform arithmetic on the elements of the equivalence classes. More specifically, we define

$$[a]_{\equiv n} + [b]_{\equiv n} = [a + b]_{\equiv n}. \tag{3.10}$$

Since there can be many elements from the same equivalence class, it is not immediately clear that this actually defines anything: what we need is that $[a_1]_{\equiv n} = [a_2]_{\equiv n}$ and $[b_1]_{\equiv n} = [b_2]_{\equiv n}$ implies $[a_1 + b_1]_{\equiv n} = [a_2 + b_2]_{\equiv n}$. Fortunately for modular arithmetic, this is true: $[a_1]_{\equiv n} = [a_2]_{\equiv n}$ means $a_1 \equiv a_2 \pmod{n}$ so n divides $a_1 - a_2$; that is, $a_1 - a_2 = r\,n$ for some whole number r. In the same way, $[b_1]_{\equiv n} = [b_2]_{\equiv n}$ means $b_1 \equiv b_2 \pmod{n}$ and $b_1 - b_2 = s\,n$ for some whole number s. Then $(a_1 + b_1) - (a_2 + b_2) = (a_1 - a_2) + (b_1 - b_2) = r\,n + s\,n = (r+s)n$ so that n divides $(a_1 + b_1) - (a_2 + b_2)$. That is, $a_1 + b_1 \equiv a_2 + b_2 \pmod{n}$ and so $[a_1 + b_1]_{\equiv n} = [a_2 + b_2]_{\equiv n}$, as we wanted. This means that addition of equivalence classes modulo n is well-defined by (3.10).

In the same way we can define multiplication of equivalence classes modulo n:

$$[a]_{\equiv n} \cdot [b]_{\equiv n} = [a \cdot b]_{\equiv n}. \tag{3.11}$$

Proving that this is well-defined is Exercise 3.11.

If we are interested in questions of divisibility, then n divides a if and only if $a \equiv 0 \pmod{n}$, which is equivalent to $[a]_{\equiv n} = [0]_{\equiv n}$.

3.6.1.3 *Separating out "large" and "small"*

Newton's method is an iterative method for approximating solutions of an equation $f(x) = 0$ is given by

$$x_{n+1} = x_n - \frac{f(x_n)}{f'(x_n)}. \tag{3.12}$$

It is well-known as a fast method when it converges. Here is the theorem and proof for it:

Theorem 3.17 (Convergence of Newton's method). *If Newton's method (3.12) converges (that is, $x_n \to \widehat{x}$ as $n \to \infty$) with f'' continuous and $f'(\widehat{x}) \neq 0$, then it converges quadratically; that is,*

$$\lim_{n\to\infty} \frac{x_{n+1} - \widehat{x}}{(x_n - \widehat{x})^2} \quad \text{is finite.}$$

Proving this involves using Taylor series with remainder. Newton's method exactly solves linear equations in one step, so it is the quadratic part of the Taylor series that is most important for determining the rate of convergence. But we need to first separate the linear part from the (small) quadratic remainder.

Proof. First we show that $f(\widehat{x}) = 0$ so $\widehat{x}$ truly is a zero of f. Taking limits on both sides of (3.12) gives

$$\widehat{x} = \lim_{n\to\infty} x_{n+1} = \lim_{n\to\infty} x_n - \frac{f(x_n)}{f'(x_n)}$$

$$= \lim_{n\to\infty} x_n - \frac{\lim_{n\to\infty} f(x_n)}{\lim_{n\to\infty} f'(x_n)}$$

$$= \widehat{x} - \frac{f(\widehat{x})}{f'(\widehat{x})} \qquad (\text{since } f'(\widehat{x}) \neq 0).$$

Solving this equation gives $f(\widehat{x}) = 0$.

From Taylor series with quadratic remainder,

$$f(x_n) = f(\widehat{x}) + f'(\widehat{x})(x_n - \widehat{x}) + \frac{1}{2}f''(c_n)(x_n - \widehat{x})^2$$

where c_n is some value between x_n and $\widehat{x}$. Also, applying Taylor series with linear remainder to f' gives

$$f'(x_n) = f'(\widehat{x}) + f''(d_n)(x_n - \widehat{x})$$

where d_n is some value between x_n and $\widehat{x}$.

We introduce the notation $e_k = x_k - \widehat{x}$ which is the "error" in x_k; our assumption is that $x_k \to \widehat{x}$ as $k \to \infty$, so that $e_k \to 0$ as $k \to \infty$. That is, e_k is small for large k. Using this notation instead of $x_k - \widehat{x}$ helps us focus on what is small and what is not.

Let $e_k = x_k - \widehat{x}$ for all k. Then

$$e_{n+1} = x_{n+1} - \widehat{x} = x_n - \widehat{x} - \frac{f(x_n)}{f'(x_n)} = e_n - \frac{f(\widehat{x} + e_n)}{f'(\widehat{x} + e_n)}$$

$$= e_n - \frac{f'(\widehat{x})e_n + \frac{1}{2}f''(c_n)e_n^2}{f'(\widehat{x}) + f''(d_n)e_n} \qquad (c_n \text{ and } d_n \text{ between } \widehat{x} \text{ and } x_n)$$

$$= e_n \left[1 - \frac{f'(\widehat{x}) + \frac{1}{2}f''(c_n)e_n}{f'(\widehat{x}) + f''(d_n)e_n} \right]$$

$$= e_n \frac{f'(\widehat{x}) + f''(d_n)e_n - f'(\widehat{x}) - \frac{1}{2}f''(c_n)e_n}{f'(\widehat{x}) + f''(d_n)e_n}$$

$$= e_n \frac{\left[f''(d_n) - \frac{1}{2}f''(c_n) \right] e_n}{f'(\widehat{x}) + f''(d_n)e_n}$$

$$= e_n^2 \frac{f''(d_n) - \frac{1}{2}f''(c_n)}{f'(\widehat{x}) + f''(d_n)e_n}.$$

Now the assumption that $x_n \to \widehat{x}$ as $n \to \infty$ means that $e_n = x_n - \widehat{x} \to 0$ as $n \to \infty$. Also since both c_n and d_n lie between x_n and $\widehat{x}$, by the squeeze principle, $c_n, d_n \to \widehat{x}$ as $n \to \infty$. As f'' is continuous, so is f'. So we can take limits of e_{n+1}/e_n^2:

$$\lim_{n \to \infty} \frac{e_{n+1}}{e_n^2} = \lim_{n \to \infty} \frac{f''(d_n) - \frac{1}{2}f''(c_n)}{f'(\widehat{x}) + f'(d_n)e_n}$$

$$= \frac{f''(\widehat{x}) - \frac{1}{2}f''(\widehat{x})}{f'(\widehat{x}) + f''(\widehat{x})\, 0} \qquad \text{(by standard rules for limits)}$$

$$= \frac{1}{2}\frac{f''(\widehat{x})}{f'(\widehat{x})},$$

which is a finite quantity, as we wanted. $\qquad\qquad\qquad\qquad\qquad\qquad\square$

3.7 Graphs and networks

In this section we return to the problem of Eulerian paths first discussed in Section 1.6.2. The definitions that we use here can be found on page 15. More information about graphs can be found in Bollobás (1979).

First we recall the statement of the theorem:

Theorem 3.18 (Euler paths or cycles). *Suppose we have a connected undirected graph. If the degree of every node is even, there is an Euler cycle; if the degree of all but two nodes is even, there is an Euler path that starts and ends at the two nodes of odd degree. If the number of nodes with odd degree is neither zero nor two, then there is no Euler path (or cycle).*

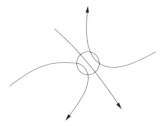

Fig. 3.2: Node in an Euler path (or cycle)

Before we start the proof, we need an idea of how to proceed. The *necessity* that the graph is connected is fairly obvious: if the graph is disconnected there are parts of the graph that cannot be reached. But the really crucial issue is whether the degree of a node is odd or even. Figure 3.2 shows a node in the middle of an Euler path (or cycle). Since it is not at the start or end of the path, the number of edges where the path comes to the node must equal the number of edges where the path leaves the node. Since every edge must be in the Euler path, the degree of the node must be even. This will show the *necessity* that the number of nodes with odd degree is zero or two.

To show *sufficiency* we need a way of creating an Euler path, or otherwise showing its existence. A natural way to do this is by means of induction on the number of edges: we add an edge to the Euler path, but then we should remove it from the graph so it is not used again. We also have to check that removing the edge does not make the graph disconnected. See, for example, the situation shown in Figure 3.3. In fact, this is the most challenging part of the entire proof.

Since we are changing the graph considered, this will change the degrees of the nodes. So our notation must take this into account. We let $\deg_G(x)$ be the degree of a node x in the graph G. Also, the induction proof that shows the existence of an Euler path or cycle needs to handle both the case of two nodes of odd degree and the case of no nodes of odd degree: it is easier to prove both together than to prove one or the other separately.

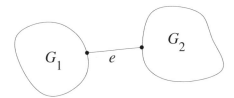

Fig. 3.3: Disconnecting a graph by removing edge e.

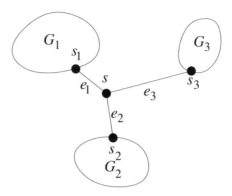

Fig. 3.4: Eulerian path: case of $\deg(s) > 1$, and removing any incident edge disconnects G

As noted above, the induction proof will involve removing edges, but we need to do so in a way that will avoid disconnecting the graph. Given a starting node s, we need to consider when we can remove an incident edge in a way that does not disconnect the graph. Sometimes, it is not possible to do this, as in Figure 3.4. So we must relate the ability to remove an edge without disconnecting the graph to the basic assumptions for Euler paths or cycles. In particular, we need to relate this ability to the number of nodes of odd degree. So we start with a Lemma about this issue:

Lemma 3.4 (Supporting lemma for Theorem 1.9). *If there is a node w of a connected graph G with $\deg_G w = m > 0$ and removing any edge incident to w disconnects G, then there are at least m nodes of G with odd degree, not counting w.*

Proof. Let $m = \deg_G w$ where w is a node G, and $e_1, e_2, \ldots, e_m$ the edges in G incident to w. Each of these edges has an endpoint different from w. (If this were not so, then an edge would form a loop, and removing that edge would not disconnect the graph.) Let x_i be the endpoint of edge e_i different from w. Removing e_i from G disconnects the graph. Let $\widehat{G}_i$ be the graph formed by all nodes connected to x_i in $G \backslash \{e_i\}$ and the corresponding edges of G.

> *We use Figure 3.4 as our guide to what we should try to prove. We will eventually show that each $\widehat{G}_i$ contains a node x where $\deg_G x$ is odd.*

First we show that $V(\widehat{G}_i) \cap V(\widehat{G}_j) = \emptyset$ if $i \neq j$. If this were not so, then there would be a node $z \in V(\widehat{G}_i) \cap V(\widehat{G}_j)$. Then there must be a path from z

to x_j and from z to x_i by definition of $\widehat{G}_j$ and $\widehat{G}_i$. Suppose that $y \in V(\widehat{G}_i)$. Then by definition of $\widehat{G}_i$ there must be a path from x_i to y; thus there is a path from y to x_j. Since e_j connects x_j to w, there is a path from y to w in $G \backslash \{e_i\}$. As G is connected, w is connected to every node in $\widehat{G}_j$ for $j \neq i$. Since we have shown that every node in $\widehat{G}_i$ is connected to w in the graph $G \backslash \{e_i\}$, then we have shown that every node in G is connected to w in $G \backslash \{e_i\}$. Thus removing e_i does not disconnect the graph G, which is a contradiction. Thus $V(\widehat{G}_i) \cap V(\widehat{G}_j) = \emptyset$ if $i \neq j$.

Now that we have shown the $\widehat{G}_i$'s to be separate, we can start looking at how many vertices have odd degree. The key is Theorem 3.9.

Theorem 3.9 says that $\sum_{x \in V(\widehat{G}_i)} \deg_{\widehat{G}_i}(x)$ is even. If $x \in V(\widehat{G}_i)$ and $x \neq x_i$ then $\deg_{\widehat{G}_i}(x) = \deg_G(x)$, but $\deg_{\widehat{G}_i}(x_i) = \deg_G(x_i) - 1$. Thus $\sum_{x \in V(\widehat{G}_i)} \deg_G(x)$ is odd. Therefore, there is an odd number of nodes $x \in V(\widehat{G}_i)$ with $\deg_G(x)$ odd; in particular, there must be at least one of them. So each $\widehat{G}_i$ must contain at least one node x with $\deg_G(x)$ odd.

The total number of nodes x in G, not counting w, with $\deg_G(x)$ odd must then be at least m, as we wanted. $\qquad \square$

With this Lemma, we can now proceed to the main proof.

Proof of Theorem 1.9. **First suppose that G has an Eulerian path or cycle:** $n_1, e_1, n_2, e_2, \ldots, n_{k-1}, e_{k-1}, n_k$. By definition, every edge e in G must be e_j for exactly one j. If n is a node in G, then unless the degree of n is zero, $n = n_j$ for various values of j. For each occurrence where $j \neq 1$ or k there is the segment e_{j-1}, n_j, e_j from the path. Each occurrence adds two to the degree of n since each edge occurs exactly once in the path. Thus if $n \neq n_1, n_k$ the degree of n must be even.

If $n_1 \neq n_k$ then we can see from the initial segment n_1, e_1 that the degree of n_1 must be odd, and from the final segment e_{k-1}, n_k that the degree of n_k must also be odd. That is there are exactly two nodes of odd degree.

If $n_1 = n_k$, then the two segments n_1, e_1 and e_{k-1}, n_k add two to the degree of $n_1 = n_k$, and so no nodes of G have odd degree.

Thus if G has an Eulerian path, then it either has exactly two nodes of odd degree (which are the start and final nodes of the Eulerian path), or no nodes of odd degree (in which case the Eulerian path is an Eulerian cycle).

Suppose that G is a connected graph with exactly two nodes of odd degree or all nodes of even degree. We prove the existence of an Eulerian path between these two nodes by induction on the number of edges of G. We designate one of the two nodes of odd degree the "start" node s, and the other the "final" node f.

Base case: If G has one edge and is connected, then $G = \bullet\!\!-\!\!-\!\!\bullet$. Clearly there is an Euler path between the two nodes of odd degree.

Suppose true for any graph with $n = k \geq 1$ edges. Show true for a graph G **with $n = k + 1$ edges.** Either there are exactly two nodes of odd degree, or no nodes of odd degree. If there are no nodes of odd degree, then we pick s (the start node) to be any node of G. Otherwise, the start node s has odd degree.

We consider the two cases: $\deg_G(s) = 1$ or $\deg_G(s) > 1$.

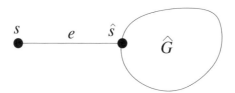

Fig. 3.5: Eulerian path: case of $\deg(s) = 1$

Case 1: If $\deg(s) = 1$ then the graph must have the form shown in Figure 3.5. Let e be the unique edge incident to s. Let $\widehat{G}$ be the graph G with node s and edge e removed. Let f be the unique node in $\widehat{G}$ with $\deg_G f$ odd. If $\widehat{s}$ is the other end-point of the edge e, then either $\widehat{s} = f$ or $\widehat{s} \neq f$. If $\widehat{s} = f$ then we have two cases to consider: $\deg_G(f) = 1$ or $\deg_G(f) > 1$. If $\deg_G(f) = 1$, then since G is connected, $G = \bullet\!\!-\!\!-\!\!\bullet$, and we are already done. If $\deg(f) > 1$, then $\widehat{G}$ has at least one edge. In this case, $\deg_{\widehat{G}}(f) = \deg_G(f) - 1$, and $\deg_{\widehat{G}}(x) = \deg_G(x)$ for any other node x in $\widehat{G}$. Thus the degree of every node in $\widehat{G}$ is even, and $\widehat{G}$ has one less edge than G. By the induction hypothesis, there is an Eulerian cycle $\widehat{s} = f, \widehat{e}_1, \widehat{n}_2, \ldots, \widehat{e}_k, f$ in $\widehat{G}$. Thus $s, e, \widehat{s}, \widehat{e}_1, \widehat{n}_2, \ldots, \widehat{e}_k, f$ is an Eulerian path in G.

If $\widehat{s} \neq f$, then note that $\deg_{\widehat{G}}(\widehat{s}) = \deg_G(\widehat{s}) - 1$, so $\deg_{\widehat{G}}(\widehat{s})$ is odd. As $\deg_{\widehat{G}}(x) = \deg_G(x)$ for all other nodes x in $\widehat{G}$, $\widehat{G}$ has exactly two nodes of odd degree: $\widehat{s}$ and f. As $\widehat{G}$ has one less edge than G, there is an Eulerian path $\widehat{s}, \widehat{e}_1, \widehat{n}_2, \ldots, \widehat{e}_k, f$ from $\widehat{s}$ to f in $\widehat{G}$. Therefore, $s, e, \widehat{s}, \widehat{e}_1, \widehat{n}_2, \ldots, \widehat{e}_k, f$ is an Eulerian path in G.

Case 2: Now consider the case $\deg_G(s) > 1$. Let $m = \deg_G(s)$ and $e_1, e_2, \ldots, e_m$ the edges incident to s. Suppose that removing *any* of the edges incident to s disconnects the graph. Then by Lemma 3.4, there must be at least m nodes x of G where $x \neq s$ and $\deg_G(x)$ is odd. If $\deg_G(s) = 2$ then at least two nodes of G have odd degree; but we choose s where $\deg_G(s)$ is even only if there are *no* nodes of odd degree in G. Thus this case is impossible. Alternatively, $\deg_G(s) > 2$ implies at least three nodes of odd degree, which contradicts our assumptions on G. Thus at least one of the edges e_1, e_2, ..., e_m can be removed without disconnecting G.

Pick e_k so that $\widehat{G} := G \backslash \{e_k\}$ is connected. Let $\widehat{s}$ be the other endpoint of e_k.

If $s = \widehat{s}$ then $\deg_{\widehat{G}}(s) = \deg_G(s) - 2$ and the number of nodes of odd degree in $\widehat{G}$ is the same as the number of nodes in G of odd degree. In particular, the parity of $\deg_G(s)$ is the same as the parity of $\deg_{\widehat{G}}(s)$.

We now assume that $\widehat{s} \neq s$. Then $\deg_{\widehat{G}}(s) = \deg_G(s) - 1$ and $\deg_{\widehat{G}}(\widehat{s}) = \deg_G(\widehat{s}) - 1$. Thus the parity of the degrees of s and $\widehat{s}$ are both changed going from G to $\widehat{G}$. We now deal with the possibilities in a case-by-case approach:

Case 2a: Suppose $\deg_G(s)$ and $\deg_G(\widehat{s})$ are both even. Then $\deg_{\widehat{G}}(s)$ and $\deg_{\widehat{G}}(\widehat{s})$ are both odd, while $\deg_G(x) = \deg_{\widehat{G}}(x)$ for all $x \neq s$, $\widehat{s}$. Since we chose $\deg_G(s) = 2$, no other node $x \neq s$, $\widehat{s}$ has odd degree either in G or $\widehat{G}$. By the induction hypothesis, there is an Euler path s, $e^{(1)}$, n_2, $e^{(2)}$, ..., $e^{(n)}$, $\widehat{s}$ in $\widehat{G}$. Then s, $e^{(1)}$, n_2, $e^{(2)}$, ..., $e^{(n)}$, $\widehat{s}$, e_k, s is an Euler cycle in G.

Case 2b: Suppose $\deg_G(s)$ is even and $\deg_G(\widehat{s})$ is odd. This cannot occur since we choose s with $\deg_G(s)$ even only if there are no nodes with odd degree in G.

Case 2c: Suppose $\deg_G(s)$ is odd and $\deg_G(\widehat{s})$ is even. Then $\deg_{\widehat{G}}(s)$ is even and $\deg_{\widehat{G}}(\widehat{s})$ is odd. Let $f \neq s$, $\widehat{s}$ be the other node in G with odd degree. Then by the induction hypothesis, there is an Euler path $\widehat{s}$, $e^{(1)}$, n_2, $e^{(2)}$, ..., f in $\widehat{G}$. Then s, e_k, $\widehat{s}$, $e^{(1)}$, n_2, $e^{(2)}$, ..., f is an Euler path in G.

Case 2d: Suppose $\deg_G(s)$ and $\deg_G(\widehat{s})$ are both odd. Then both $\deg_{\widehat{G}}(s)$ and $\deg_{\widehat{G}}(\widehat{s})$ are even. By the induction hypothesis there is an Euler cycle $\widehat{s}$, $e^{(1)}$, n_2, $e^{(2)}$, ..., $\widehat{s}$ in $\widehat{G}$. Then s, e_k, $\widehat{s}$, $e^{(1)}$, n_2, $e^{(2)}$, ..., $\widehat{s}$ is an Euler path in G.

In all cases, G has an Euler cycle or an Euler path, as we wanted. $\qquad\square$

It is possible to prove, for example, that if G is connected and all nodes have even degree then G has an Euler cycle directly, without showing the above result for exactly two nodes of odd degree. But to do this, two edges have to be considered in each step: the "surgery" that needs to be done involves replacing two edges with one. Again, care has to be taken to avoid or handle disconnecting the graph into two parts.

3.8 Real numbers and convergence

Real numbers are used whenever we deal with "continuous" problems, so we need to understand the rules that apply to real numbers. The usual algebraic properties that hold for rational numbers (like $a(b+c) = ab+ac$) also apply to real numbers. What makes real numbers different involves infinite processes, such as limits. Often the essential property of real numbers is described as the Least Upper Bound property. An equivalent property is that all bounded increasing sequences of real numbers converge. Both of these are described in the next section.

3.8.1 *The Least Upper Bound property*

In dealing with convergence, we have to grapple with the nature of real numbers. This is often expressed by the *Least Upper Bound property* for real numbers:

> If S is a set of real numbers that is bounded above (that is, there is an M where $x \in S \Rightarrow x \le M$), then S has a least upper bound, being a real number L where $x \in S \Rightarrow x \le L$ and for any real number L' where $x \in S \Rightarrow x \le L'$ we have $L \le L'$.

The number L is called the *supremum* of S, and is denoted $\sup S$.

The Least Upper Bound property is typically given as an axiom in many calculus textbooks. This property is often given as the defining property that real numbers have, but rational numbers do not. Some refer to this property as the *completeness axiom*. This is equivalent to the claim that *all bounded increasing sequences converge*. (Proving that convergence of bounded increasing sequences implies the Least Upper Bound property is Exercise 3.19.) Convergence of a sequence a_k, $k = 1, 2, 3, \ldots$ to a limit a^* means that for any $\epsilon > 0$ there is an N where $n \ge N \Rightarrow |a_n - a^*| < \epsilon$. We usually write $\lim_{k \to \infty} a_k = a^*$ or $a_k \to a^*$ as $k \to \infty$. Note that if a sequence a_k does converge, there can only be one limit: if a^* and $\widehat{a}$ were two limits, then given $\epsilon > 0$ we have an N where $n \ge N \Rightarrow |a^* - \widehat{a}| \le |a^* - a_n| + |a_n - \widehat{a}| < \epsilon + \epsilon = 2\epsilon$. We can make $\epsilon > 0$ as small as we please, so $|a^* - \widehat{a}| = 0$ and therefore $a^* = \widehat{a}$.

We would like to prove this instead of treating it as an axiom. The problem is that we need to understand what real numbers are first. One way, is as infinite decimal expansions, with an optional minus sign for negative numbers. Of course, we have to accept that $0.999999999\ldots = 1.00000000\ldots$; if you are not convinced, then try working out the difference, which you will find is zero.

First we can prove that

Theorem 3.19. *All bounded increasing sequences of real numbers converge.*

Think about strategy for a moment. We need to build the proof around the definition of real numbers. If we define them in terms of infinitely long strings of digits with a single decimal point, then we need to prove the result in terms of these strings of digits.

Proof. Suppose we have a bounded increasing sequence a_k, $k = 1, 2, 3, \ldots$ that is bounded above by M. Then for all k, $a_1 < a_k < a_{k+1} \leq M$. Set

$$b_k = \frac{a_k - a_1}{M - a_1},$$

so that $0 \leq b_k < 1$ for all k. Note that $b_{k+1} = (a_{k+1} - a_1)/(M - a_1) > (a_k - a_1)/(M - a_1) = b_k$ (since $M - a_1 > 0$) so the sequence b_k is also an increasing bounded sequence.

Write out each b_k as an infinite decimal:

$$b_k = 0.b_k^{(1)} b_k^{(2)} b_k^{(3)} \cdots$$

where each $b_k^{(j)}$ is a digit (0, 1, 2, 3, 4, 5, 6, 7, 8, 9), but we do not allow the decimal to end with 9 recurring.

Since $b_{k+1} \geq b_k$, we must have the first digit non-decreasing: $b_{k+1}^{(1)} \geq b_k^{(1)}$. Since there are only ten possible choices for $b_k^{(1)}$, as k increases $b_k^{(1)}$ must become fixed. So there must be a K_1 where $k \geq K_1 \Rightarrow b_k^{(1)} = c^{(1)}$.

Since $b_{k+1} \geq b_k$, for $k \geq K_1$ we have $b_{k+1}^{(2)} \geq b_k^{(2)}$. Again, the sequence $b_k^{(2)}$ must eventually be constant as there are only the same ten possibilities for $b_k^{(2)}$. Let K_2 be chosen so that $K_2 \geq K_1$ and $k \geq K_2 \Rightarrow b_{k+1}^{(2)} = b_k^{(2)} = c^{(2)}$.

Continuing in this way, suppose we have chosen K_j so that $k \geq K_j$ implies $b_{k+1}^{(r)} = b_k^{(r)}$ for $r = 1, 2, \ldots, j$. Then $b_{k+1} \geq b_k$ for all k implies that for $k \geq K_j$ we have $b_{k+1}^{(j+1)} \geq b_k^{(j+1)}$. Since there are only ten possible values for $b_k^{(j+1)}$, $b_k^{(j+1)}$ must eventually be constant as $k \to \infty$. Choose K_{j+1} so that $K_{j+1} \geq K_j$ and $k \geq K_{j+1} \Rightarrow b_{k+1}^{(j+1)} = b_k^{(j+1)}$. Let $c^{(j+1)} = b_k^{(j+1)}$ for $k \geq K_{j+1}$.

By induction on j, for each j there is a K_j where $k \geq K_j$ implies $b_{k+1}^{(j)} = b_k^{(j)}$. Let $c^{(j)} = \lim_{k \to \infty} b_k^{(j)}$ and $c = 0.c^{(1)} c^{(2)} c^{(3)} \cdots$. We now show that $b_k \to c$ as $k \to \infty$. Let $\epsilon > 0$ be given. Then there is a j where $0 < 10^{-j} < \epsilon$. Then for $k \geq K_j$ we have $b_k^{(r)} = c^{(r)}$ for $r = 1, 2, \ldots, j$, and so $|b_k - c| \leq 10^{-j} < \epsilon$ as we wanted. Thus every non-decreasing bounded sequence has a limit. $\qquad \square$

Note that most of the hard work goes into constructing the limit c. As part of this construction we prove the important point that the jth digits of b_k are eventually constant as $k \to \infty$.

As a side note, although we assume that the decimal expansions of b_k do not end in $\ldots 99999999 \ldots$, there is no guarantee that this will also be true of the decimal expansion of c. In fact, if the sequence b_k is strictly increasing ($b_{k+1} > b_k$) and $\lim_{k \to \infty} b_k = 1$, then $c^{(j)} = 9$ for all j.

3.8.2 *Convergence of infinite sums*

Suppose we have an infinite series $\sum_{k=0}^{\infty} a_k(r)$ and $a_k(r) \to a_k^*$ as $r \to r^*$. Are we justified in saying that $\sum_{k=0}^{\infty} a_k^* = \lim_{r \to r^*} \sum_{k=0}^{\infty} a_k(r)$? For finite sums we have $\sum_{k=0}^{N} a_k^* = \lim_{r \to r^*} \sum_{k=0}^{N} a_k(r)$, but in general for infinite sums, the answer is no. Here is a counterexample: let $a_k(r) = (r^k/k!)e^{-r}$. Then

$$\sum_{k=0}^{\infty} a_k(r) = \sum_{k=0}^{\infty} \frac{r^k}{k!} e^{-r} = e^r e^{-r} = 1 \to 1 \qquad \text{as } r \to \infty.$$

On the other hand, $a_k^* = \lim_{r \to \infty} a_k(r) = \lim_{r \to \infty} r^k/(k! \, e^r) = 0$ from the general principle that exponentials dominate polynomials. (L'Hospital's rule can also be used to prove this.) Thus $\sum_{k=0}^{\infty} a_k^* = 0 \neq 1 = \lim_{r \to \infty} \sum_{k=0}^{\infty} a_k(r)$.

We need something else than "$a_k(r) \to a_k^*$ as $r \to r^*$" to seal the argument. Here is a theorem that supplies that "something else".

Theorem 3.20. *Suppose that $a_k(r) \to a_k^*$ as $r \to r^*$ for all k and that $\sum_{k=0}^{\infty} a_k(r)$ converges and is finite for all r. Assume also that $|a_k(r)| \leq b_k$ for all k and r, and that $\sum_{k=0}^{\infty} b_k$ converges (and so is finite). Then*

$$\sum_{k=0}^{\infty} a_k^* = \lim_{r \to r^*} \sum_{k=0}^{\infty} a_k(r).$$

Before we start, we note again that finite sums converge; for the (infinite) remainder we use the bound of b_k.

In the middle of the proof we use a *subtract and add* move to help bound the errors: we use

$$a - b = (a - c) + (c - b)$$

where c is chosen so that we can estimate $a - c$ and $c - b$. Actually we go a little further: we use

$$a - b = (a - c) + (c - d) + (d - b).$$

We just need to choose c and d so we can estimate the differences.

Proof. We want to show that for any $\epsilon > 0$ there is a $\delta > 0$ where $|r - r^*| < \delta$ implies $\left| \sum_{k=0}^{\infty} a_k(r) - \sum_{k=0}^{\infty} a_k^* \right| < \epsilon$. For any finite N,

$$\lim_{r \to r^*} \sum_{k=0}^{N} a_k(r) = \sum_{k=0}^{N} \lim_{r \to r^*} a_k(r) = \sum_{k=0}^{N} a_k^*.$$

We want to split the error into two parts: the error in the finite part of the sum, and the error in the infinite remainder. Let us aim to get each of these less than $\epsilon/4$ so that the total error is less than ϵ. (We will allow extra wiggle room for extra problems that come up.) The hardest part is to bound the infinite remainder.

Now $\sum_{k=0}^{\infty} b_k$ converges to some value b^*, so $\lim_{N \to \infty} \sum_{k=0}^{N} b_k = b^*$. But all b_k's are non-negative, so if $M \geq N$, then $b^* \geq \sum_{k=0}^{M} b_k \geq \sum_{k=0}^{N} b_k$. Therefore, $b^* - \sum_{k=0}^{N} b_k \geq \sum_{k=N+1}^{M} b_k$ for all $M \geq N$. We can make $b^* - \sum_{k=0}^{N} b_k$ as close to zero as we please by making N sufficiently large. Choose N large enough so that $b^* - \sum_{k=0}^{N} b_k < \epsilon/8$. Then $\sum_{k=N+1}^{M} b_k < \epsilon/8$ for any $M \geq N$. Taking $M \to \infty$ gives $\sum_{k=N+1}^{\infty} b_k \leq \epsilon/8$.

Once N is chosen, we can choose δ.

Choose $\delta > 0$ so that $|r - r^*| < \delta$ implies $\left| \sum_{k=0}^{N} a_k(r) - \sum_{k=0}^{N} a_k^* \right| < \epsilon/4$.

We need to approximate the infinite sums by finite sums. Here we need some of that extra wiggle room.

Since both series $\sum_{k=0}^{\infty} a_k(r)$ and $\sum_{k=0}^{\infty} a_k^*$ converge, there is an $M \geq N$ where

$$\left| \sum_{k=0}^{M} a_k(r) - \sum_{k=0}^{\infty} a_k(r) \right| = \left| \sum_{k=M+1}^{\infty} a_k(r) \right| \leq \sum_{k=M+1}^{\infty} b_k \leq \epsilon/8,$$

$$\left| \sum_{k=0}^{M} a_k^* - \sum_{k=0}^{\infty} a_k^* \right| < \epsilon/8.$$

Then for $|r - r^*| < \delta$ we have

$$
\left| \sum_{k=0}^{\infty} a_k(r) - \sum_{k=0}^{\infty} a_k^* \right|
$$

$$
= \left| \sum_{k=0}^{\infty} a_k(r) - \sum_{k=0}^{M} a_k(r) + \sum_{k=0}^{M} a_k(r) - \sum_{k=0}^{M} a_k^* + \sum_{k=0}^{M} a_k^* - \sum_{k=0}^{\infty} a_k^* \right|
$$

$$
< \left| \sum_{k=0}^{M} a_k(r) - \sum_{k=0}^{M} a_k^* \right| + \frac{\epsilon}{8} + \frac{\epsilon}{8}
$$

$$
\leq \left| \sum_{k=0}^{N} a_k(r) - \sum_{k=0}^{N} a_k^* \right| + \left| \sum_{k=N+1}^{M} a_k(r) - \sum_{k=N+1}^{M} a_k^* \right| + \frac{\epsilon}{4}
$$

$$
\leq \frac{\epsilon}{4} + \sum_{k=N+1}^{M} |a_k(r) - a_k^*| + \frac{\epsilon}{4}
$$

$$
\leq \frac{\epsilon}{2} + \sum_{k=N+1}^{M} (|a_k(r)| + |a_k^*|)
$$

$$
\leq \frac{\epsilon}{2} + \sum_{k=N+1}^{M} 2b_k \leq \frac{\epsilon}{2} + 2 \times \frac{\epsilon}{4} = \epsilon,
$$

as we wanted. □

We can only really find out that we needed to use $\epsilon/8$ (or maybe something even smaller) as we go through the proof. You can write a "draft" proof where you use some other symbol for the bounds used at the different stages: $\sum_{k=N+1}^{M} b_k < \eta$ for $M \geq N$, $\left| \sum_{k=0}^{N} a_k(r) - \sum_{k=0}^{N} a_k^* \right| < \eta$, $\left| \sum_{k=0}^{M} a_k^* - \sum_{k=0}^{\infty} a_k^* \right| < \eta$. Then the final inequality might be: for $|r - r^*| < \delta$,

$$
\left| \sum_{k=0}^{\infty} a_k(r) - \sum_{k=0}^{\infty} a_k^* \right| < \cdots \leq 10\eta.
$$

Then go back to just after the introduction of ϵ and put "*Let $\eta = \epsilon/10$.*"

3.8.3 *Different kinds of convergence*

When we are dealing with functions, say $[a, b] \to \mathbb{R}$, we can have many different ideas for when a sequence of functions f_n, $n = 1, 2, 3, \ldots$ converges, such as:

- *pointwise convergence*: $f_n(x) \to f(x)$ as $n \to \infty$ for all (or almost all) $x \in [a, b]$.

- *uniform convergence*: $\max_{a \le x \le b} |f_n(x) - f(x)| \to 0$ or $\sup_{a \le x \le b} |f_n(x) - f(x)| \to 0$ as $n \to \infty$.
- L^2 *convergence*: $\int_a^b (f_n(x) - f(x))^2 \, dx \to 0$ as $n \to \infty$.
- L^1 *convergence*: $\int_a^b |f_n(x) - f(x)| \, dx \to 0$ as $n \to \infty$.
- *weak* convergence*: $\int_a^b f_n(x) g(x) \, dx \to \int_a^b f(x) g(x) \, dx$ as $n \to \infty$ for any continuous $g \colon [a, b] \to \mathbb{R}$.

Each of these notions of convergence is different and has its own uses. Selecting which one is appropriate is an important part of developing a proof, or a theorem. Pointwise convergence is often the easiest condition to check, but uniform convergence is often the easiest to use. To see the difference between them, consider the sequence $f_n(x) = x^n$, $f_n \colon [0, 1] \to \mathbb{R}$ for $n = 1, 2, 3, \ldots$. Then for $0 \le x < 1$, $f_n(x) \to 0$ as $n \to \infty$, and $f_n(1) = 1$ for all n. So $f_n \to f$ pointwise where $f(x) = 0$ for $0 \le x < 1$ and $f(1) = 1$. On the other hand, $\sup_{0 \le x \le 1} |f_n(x) - f(x)| = \sup_{0 \le x < 1} x^n = 1$ for all n so $f_n \not\to f$ uniformly.

Using uniform convergence, we can prove important results, such as:

Theorem 3.21. *If* $f_n' \to f'$ *uniformly on* $[a, b]$ *with* $f_n(a) \to f(a)$, *then* $f_n \to f$ *uniformly on* $[a, b]$.

Proof. Suppose $f_n' \to f'$ uniformly on $[a, b]$ and $f_n(a) \to f(a)$. We want to show that $\max_{a \le x \le b} |f_n(x) - f(x)| \to 0$ as $n \to \infty$.

> *We need to use the information about* f_n' *to give us information about* f_n, *so we should use the fundamental theorem of calculus.*

Now $f_n(x) = f_n(a) + \int_a^x f_n'(s) \, ds$ and $f(x) = f(a) + \int_a^x f'(s) \, ds$, so

$$|f_n(x) - f(x)| = \left| f_n(a) - f(a) + \int_a^x f_n'(s) \, dx - \int_a^x f'(s) \, ds \right|$$

$$\le |f_n(a) - f(a)| + \left| \int_a^x (f_n'(s) - f'(s)) \, dx \right|$$

$$\le |f_n(a) - f(a)| + \int_a^x |f_n'(s) - f'(s)| \, ds.$$

For $a \le s \le b$, $|f_n'(s) - f'(s)| \le \max_{a \le r \le b} |f_n'(r) - f'(r)|$. So for $a \le x \le b$,

$$|f_n(x) - f(x)| \le |f_n(a) - f(a)| + \int_a^x \max_{a \le r \le b} |f_n'(r) - f'(r)| \, ds$$

$$= |f_n(a) - f(a)| + (x - a) \max_{a \le r \le b} |f_n'(r) - f'(r)|$$

$$\le |f_n(a) - f(a)| + (b - a) \max_{a \le r \le b} |f_n'(r) - f'(r)|.$$

The right-hand side now does not depend on x, so

$$\max_{a \leq x \leq b} |f_n(x) - f(x)| \leq |f_n(a) - f(a)| + (b-a) \max_{a \leq r \leq b} |f_n'(r) - f'(r)|.$$

Now $|f_n(a) - f(a)| \to 0$ as $n \to \infty$, and $\max_{a \leq r \leq b} |f_n'(r) - f'(r)| \to 0$ as $n \to \infty$, so $\max_{a \leq x \leq b} |f_n(x) - f(x)| \to 0$ as $n \to \infty$, as we wanted. $\qquad \square$

3.8.4 *"Give an ϵ of room... "*

This is a favorite saying of Terrence Tao, a famous mathematician[1] at UCLA. Very often we want to prove an equality or inequality for real numbers, but we cannot do this directly. Instead, we allow for errors up to a certain error tolerance, and show that the error tolerance can be as small as we please. Since we usually use $\epsilon > 0$ to represent a suitable error tolerance, we "give an ϵ of room...."

This approach is particularly useful in dealing with limiting values or infinite objects. For example, suprema and infima (see Exercise 3.19) are often best handled using this kind of argument.

Theorem 3.22. *Let $f_\alpha \colon \mathbb{R}^n \to \mathbb{R}$ for $\alpha \in J$ be a (possibly infinite) family of convex functions and $g(\mathbf{z}) := \sup_{\alpha \in J} f_\alpha(\mathbf{z})$ has a finite value for all $\mathbf{z}$. Then g is also a convex function.*

Proof. Let $\mathbf{x}$ and $\mathbf{y}$ be two points in $\mathbb{R}^n$ and $0 \leq \theta \leq 1$. We want to show that $g(\theta \mathbf{x} + (1-\theta)\mathbf{y}) \leq \theta \, g(\mathbf{x}) + (1-\theta) g(\mathbf{y})$.

> *We have a clear statement of purpose. But what we will actually prove is that for any $\epsilon > 0$ we have $g(\theta \mathbf{x} + (1-\theta)\mathbf{y}) \leq \theta \, g(\mathbf{x}) + (1-\theta)g(\mathbf{y}) + \epsilon$.*

Let $\epsilon > 0$ be given. There must be an $\alpha \in J$ where $g(\theta \mathbf{x} + (1-\theta)\mathbf{y}) \leq f_\alpha(\theta \mathbf{x} + (1-\theta)\mathbf{y}) + \epsilon$. But f_α is convex, so

$$f_\alpha(\theta \mathbf{x} + (1-\theta)\mathbf{y}) \leq \theta \, f_\alpha(\mathbf{x}) + (1-\theta) \, f_\alpha(\mathbf{y}) \qquad \text{and thus}$$

$$g(\theta \mathbf{x} + (1-\theta)\mathbf{y}) \leq \theta \, f_\alpha(\mathbf{x}) + (1-\theta) \, f_\alpha(\mathbf{y}) + \epsilon.$$

Now $f_\alpha(\mathbf{x}) \leq g(\mathbf{x})$ and $f_\alpha(\mathbf{y}) \leq g(\mathbf{y})$ from the definition of g. Since θ, $1-\theta \geq 0$,

$$g(\theta \mathbf{x} + (1-\theta)\mathbf{y}) \leq \theta \, g(\mathbf{x}) + (1-\theta) \, g(\mathbf{y}) + \epsilon.$$

As this is true for any $\epsilon > 0$, we have $g(\theta \mathbf{x} + (1-\theta)\mathbf{y}) \leq \theta \, g(\mathbf{x}) + (1-\theta) \, g(\mathbf{y})$, as we wanted. $\qquad \square$

---

[1]Famous for a mathematician is not like being Albert Einstein or winning a Nobel Prize — there is no Nobel prize in mathematics. But Terrence Tao did win the Fields Medal, which is the closest thing for mathematics.

This theorem can be extended to allow $g(\mathbf{z}) = +\infty$ for some $\mathbf{z}$. In particular, if $g(\theta\mathbf{x} + (1 - \theta)\mathbf{y}) = +\infty$ then for any $\epsilon > 0$ we can choose $\alpha \in J$ so that $f_\alpha(\theta\mathbf{x} + (1-\theta)\mathbf{y}) > 1/\epsilon$. The argument can then be completed in the same spirit as the above proof.

3.9 Approximating or building "bad" things with "nice" things

Often we can prove properties of "nice" things that are hard to prove about "bad" things. But once we have the proof for "nice" things, often we can use some kind of limiting or extension argument to show that the result must be true for "bad" things as well.

Consider the power functions $f_a(x) = x^a$ for $x > 0$. To prove the usual properties of exponents for real a we start with considering a to be a positive whole number. Then x^a can be defined by repeated multiplication. From this beginning, we can extend the definition to integers (including $a = 0$ and a negative): $x^0 = 1$ and $x^{-a} = 1/x^a$. Then we can extend the definition to rational $a = r/s$: $y = x^{r/s}$ means $y^s = x^r$. Finally, if a is irrational, then we have to define x^a as a limit: $x^a = \lim_{b \to a,\, b \in \mathbb{Q}} x^b$. In this way we have gone from the "nicest" exponents (positive whole numbers) to the "nastiest" (irrational numbers). This is an example of *bootstrapping*.

Theorem 3.23. *If x, $y > 0$ then $x^{a+b} = x^a\, x^b$, $(x^a)^b = x^{ab}$ and $(xy)^a = x^a\, y^a$ for any real a and b.*

Note that in going from integers to rationals, we need the $(x^a)^b = x^{ab}$ argument (for integer a and b) to show $x^{a+b} = x^a\, x^b$ (for rational a and b). We also need $(xy)^a = x^a\, y^a$ (for rational a) to prove $(x^a)^b = x^{ab}$ (for real a and b). This is an example of where aiming to prove more rather than less makes the proof easier.

Proof. **Positive integers:** First we consider a, b to be positive integers. In this case,

$$x^{a+b} = \underbrace{x\,x\,x\,\cdots\,x}_{a \text{ times}} \underbrace{x\,x\,\cdots\,x}_{b \text{ times}} = x^a\, x^b.$$

Also,

$$(x^a)^b = \underbrace{x^a\,x^a\,\cdots\,x^a}_{b\text{ times}} = \underbrace{\underbrace{x\,x\,\cdots\,x}_{a\text{ times}}\underbrace{x\,x\,\cdots\,x}_{a\text{ times}}\cdots\underbrace{x\,x\,\cdots\,x}_{a\text{ times}}}_{b\text{ times}}$$

$$= \underbrace{x\,x\,x\,\cdots\,x}_{ab\text{ times}} = x^{ab}.$$

Finally,

$$(xy)^a = \underbrace{(xy)\,(xy)\,\cdots\,(xy)}_{a\text{ times}} = \underbrace{x\,x\,\cdots\,x}_{a\text{ times}}\underbrace{y\,y\,y\,\cdots\,y}_{a\text{ times}} = x^a\,y^a.$$

Integers: Now we consider a, b to be integers. If both are positive, then the previous arguments apply. If $b = 0$, then $x^{a+0} = x^a = x^a\,x^0$ since $x^0 = 1$. Also $(x^a)^0 = 1 = x^0 = x^{0\,a}$. If $a = 0$, then $x^{0+b} = x^b = x^0\,x^b$, again since $x^0 = 1$. Also, $(x^0)^b = 1^b = \underbrace{1\,1\,\cdots\,1}_{b\text{ times}} = 1$. Finally, $(xy)^0 = 1 = x^0\,y^0$.

If $a > 0$ and $b < 0$ then

$$x^a\,x^b = x^a\,1/x^{-b} = \underbrace{x\,x\,\cdots\,x}_{a\text{ times}}\,\frac{1}{\underbrace{x\,x\,\cdots\,x}_{-b\text{ times}}}$$

$$= \left\{\begin{array}{ll} \underbrace{x\,x\,\cdots\,x}_{a-(-b)\text{ times}} & \text{if } a > -b, \\ 1/(\underbrace{x\,x\,\cdots\,x}_{(-b)-a\text{ times}}) & \text{if } a < -b, \\ 1 & \text{if } a = -b, \end{array}\right\} = x^{a+b}.$$

Also, $(x^a)^b = 1/(x^a)^{-b} = 1/(x^{a(-b)}) = 1/(x^{-ab}) = x^{-(-ab)} = x^{ab}$.

If $a < 0$ and $b > 0$ then we can show $x^a\,x^b = x^{a+b}$ using the previous argument, since it is symmetrical in a and b. Also,

$$(x^a)^b = (1/x^{-a})^b = \underbrace{(1/x^{-a})(1/x^{-a})\cdots(1/x^{-a})}_{b\text{ times}}$$

$$= 1/(\underbrace{x^{-a}\,x^{-a}\,\cdots\,x^{-a}}_{b\text{ times}}) = 1/(x^{-a})^b = 1/(x^{-ab})$$

$$= x^{-(-ab)} = x^{ab}.$$

If $a < 0$ and $b < 0$ then $x^a\,x^b = (1/x^{-a})(1/x^{-b}) = 1/(x^{-a}\,x^{-b}) = 1/(x^{-a-b}) = x^{-(-a-b)} = x^{a+b}$, and $(x^a)^b = 1/(x^a)^{-b} = 1/(x^{a(-b)}) = x^{-a(-b)} = x^{ab}$.

Finally, for $a < 0$, $(xy)^a = 1/(xy)^{-a} = 1/(x^{-a}\,y^{-a}) = (1/x^{-a})(1/y^{-a}) = x^a\,y^a$. This covers all cases for a, b integers.

Rational numbers: Now we consider a, b to be rational numbers: $a = r/s$ and $b = t/u$ with s, $u \neq 0$ and r, s, t, and u integers. This extension now includes irrational *values* such as $2^{1/2}$. Thus all the usual rules hold for exponents r, s, t, and u.

First note that the definition is well-defined; that is, for each $x > 0$ there is exactly one value $y = x^a$: from the definition, $y = x^a$ is equivalent to $y > 0$ and $y^s = x^r$. The function $y \mapsto y^s$ ($s \neq 0$) is an increasing continuous function if $s > 0$ and a decreasing continuous function if $s < 0$. As $\lim_{y \to \infty} y^s = +\infty$, $\lim_{y \to 0+} y^s = 0$ if $s > 0$ and $\lim_{y \to \infty} y^s = 0$, $\lim_{y \to 0+} y^s = +\infty$ if $s < 0$, there is a value $y > 0$ where $y^s = x^r > 0$ by the intermediate value theorem. Since the function $y \mapsto y^s$ is either strictly increasing or strictly decreasing for $s \neq 0$, there is exactly one value y where $y^s = x^r$. However, to complete the proof that x^a is well-defined we need to show that if $r/s = r'/s'$ then $x^{r/s} = x^{r'/s'}$. That is, if $r/s = r'/s'$ and $y^s = x^r$ then $y^{s'} = x^{r'}$. We do this for the special case where $r' = r/d$ and $s' = s/d$ where d is the common factor of r and s, so r' and s' have no common factors. If $y^{s'} = x^{r'}$ then $(y^{s'})^d = (x^{r'})^d$; that is, $y^{ds'} = x^{dr'}$ or $y^s = x^r$. Conversely, if $y^{s'} \neq x^{r'}$, then since both are positive and $d > 0$, $y^s = (y^{s'})^d \neq (x^{r'})^d = x^r$. So $y^s = x^r$ if and only if $y^{s'} = x^{r'}$.

If $x^a = y_1$ and $x^b = y_2$, then $x^r = y_1^s$ and $x^t = y_2^u$ and so $(y_1 y_2)^{su} = (y_1^s)^u (y_2^u)^s = (x^r)^u (x^t)^s = x^{ru} x^{ts} = x^{ru+ts}$, and so $y_1 y_2 = x^{(ru+ts)/(su)}$. That is, $x^a x^b = y_1 y_2 = x^{(ru+ts)/(su)}$. But $(ru + ts)/su = r/s + t/u = a + b$, so $x^{a+b} = x^a x^b$.

Also, $(x^a)^b = w^b = z$ if and only if $w^t = z^u$. Since, $x^r = w^s$, $x^{rt} = (x^r)^t = (w^s)^t = w^{st} = (w^t)^s = (z^u)^s = z^{us}$. Then $z = x^{(rt)/(su)} = x^{ab}$ as $ab = \dfrac{r}{s}\dfrac{t}{u} = \dfrac{rt}{su}$. That is, $(x^a)^b = x^{ab}$.

Finally, suppose $x^a = u > 0$ and $y^a = v > 0$. Then $x^r = u^s$ and $y^r = v^s$, and so $(xy)^r = x^r y^r = u^s v^s = (uv)^s$. Therefore, $(xy)^a = uv = x^a y^a$.

Real numbers: For this we need some limiting arguments and monotonicity results: if $x > 1$ and $a > 0$ is rational, then $x^a > 1$. First we show that for $x > 1$, x^a is an increasing function of a: if $b > a$ are both rational then $x^b = x^{a+(b-a)} = x^a x^{b-a} > x^a 1 = x^a$ since $x^{b-a} > 1$ as $b - a > 0$. We want to show that x^a is a continuous function of $a \in \mathbb{Q}$; that is, given a rational a and $\epsilon > 0$ there is $\delta > 0$ where $|b - a| < \delta$ implies $\left| x^b - x^a \right| < \epsilon$ for any rational b. Note that $x^b - x^a = x^a(x^{b-a} - 1)$. Since x^{b-a} is an increasing function of $b - a$, if $|b - a| < 1/n$ for an integer n, then $-1/n < b - a < +1/n$ and $1/(x^{1/n}) = x^{-1/n} \leq x^{b-a} \leq x^{1/n}$. Now we need to bound $x^{1/n}$. Note that $y = x^{1/n}$ if and only if $y^n = x$ and $y > 0$. We see that $x > 1$ implies that $y > 1$. Writing $y = 1 + (y - 1)$, $y^n = (1 + (y - 1))^n$. Since $y - 1 > 0$, by the binomial theorem

$$x = y^n = (1 + (y-1))^n = \sum_{k=0}^{n} \binom{n}{k} 1^{n-k} (y-1)^k$$

$$> \binom{n}{0} 1^n (y-1)^0 + \binom{n}{1} 1^{n-1} (y-1)^1 = 1 + n(y-1).$$

Therefore, $0 < y - 1 < (x-1)/n$. So $1 < x^{1/n} < 1 + (x-1)/n$. Taking $n \to \infty$ shows that $\lim_{n\to\infty} x^{1/n} = 1$ by the squeeze theorem. Also $\lim_{x\to 1+} x^{1/n} = 1$ by the squeeze theorem. Similarly, $\lim_{x\to 1-} x^{1/n} = 1/(\lim_{x\to 1-} x^{-1/n}) = 1/(\lim_{x\to 1-} (1/x)^{1/n}) = 1/(\lim_{z\to 1+} z^{1/n}) = 1/1 = 1$ as $0 < x < 1$ implies $z = 1/x > 1$ and $x \to 1$ implies $z \to 1$.

For any real number α there is a sequence of rational numbers a_k where $\alpha = \lim_{k\to\infty} a_k$. We can define $x^\alpha = \lim_{k\to\infty} x^{a_k}$. First we show that the limit exists for $x > 1$. Now the infimum $z_1 := \inf \{ x^a \mid a \in \mathbb{Q}, a > \alpha \}$ exists since $\{ x^a \mid a \in \mathbb{Q}, a > \alpha \}$ is a non-empty set that is bounded below by zero. Also the supremum $z_2 := \sup \{ x^a \mid a \in \mathbb{Q}, a < \alpha \}$ exists since it is a non-empty set bounded above by x^n for any integer $n \geq \alpha$. Note that $z_2 \leq z_1$ since $a_2 < \alpha < a_1$ implies $x^{a_2} < x^{a_1}$. For any $\epsilon > 0$ there are $a_2 < \alpha < a_1$ where $0 \leq x^{a_1} - z_1 < \epsilon$ and $0 \leq z_2 - x^{a_2} < \epsilon$. Since x^a is an increasing function of a for $x > 1$, for any rational a_1', a_2' where $a_2 > a_2' > \alpha > a_1' > a_1$ we have $0 \leq x^{a_1'} - z_1 < x^{a_1} - z_1 < \epsilon$ and $0 \leq z_2 - x^{a_2'} < z_2 - x^{a_2} < \epsilon$. Adding gives $0 \leq x^{a_1'} - x^{a_2'} + z_2 - z_1 < 2\epsilon$. Now $x^{a_1'} - x^{a_2'} = x^{a_2'}(x^{a_1'-a_2'} - 1) \to 0$ if $a_1' - a_2' \to 0$. So $0 \leq z_2 - z_1$ and $z_1 \leq z_2$. As we already have $z_2 \leq z_1$, this implies that $z_1 = z_2$. This common value is understood to be x^α.

Furthermore, $0 \leq x^{a_1'} - x^{a_2'} < 2\epsilon$. Setting $\delta = \min(\alpha - a_1, a_2 - \alpha)$, if $|a - \alpha| < \delta$ and a rational, then $a_1 < a < a_2$, which in turn implies that $|x^a - x^\alpha| < 2\epsilon$. Thus $\lim_{a\to\alpha; a\in\mathbb{Q}} x^a = x^\alpha$, and if we have a sequence $a_k \to \alpha$ as $k \to \infty$ with a_k rational for all k, then $\lim_{k\to\infty} x^{a_k} = x^\alpha$.

If $0 < x < 1$, we define $x^\alpha = 1/((1/x)^\alpha) = ((x^{-1})^\alpha)^{-1}$. Then if $a_k \to \alpha$ as $k \to \infty$ with $a_k \in \mathbb{Q}$ for all k, we have

$$\lim_{k\to\infty} x^{a_k} = \lim_{k\to\infty} ((x^{-1})^{a_k})^{-1} = (\lim_{k\to\infty} (x^{-1})^{a_k})^{-1} = ((x^{-1})^\alpha)^{-1} = x^\alpha,$$

as desired.

Now consider, in addition, a sequence $b_k \to \beta \in \mathbb{R}$ as $k \to \infty$ with $b_k \in \mathbb{Q}$ for all k. The first and last rules of exponents for real exponents are then justified by the following calculations:

$$x^{\alpha+\beta} = \lim_{k\to\infty} x^{a_k+b_k} = \lim_{k\to\infty} x^{a_k} x^{b_k} = \lim_{k\to\infty} x^{a_k} \lim_{k\to\infty} x^{b_k} = x^\alpha x^\beta;$$

$$(xy)^\alpha = \lim_{k\to\infty} (xy)^{a_k} = \lim_{k\to\infty} x^{a_k} y^{a_k} = \lim_{k\to\infty} x^{a_k} \lim_{k\to\infty} y^{a_k} = x^\alpha y^\alpha.$$

For showing $(x^\alpha)^\beta = x^{\alpha\beta}$ we need to show that if $x > 1$, γ is a *real* number and $|\gamma| \le 1/n$ with n a positive integer, we have $x^{-2/n} \le x^\gamma \le x^{+2/n}$. Suppose that $c_k \to \gamma$ as $k \to \infty$ with each $c_k \in \mathbb{Q}$. Then $|c_k| \le 2/n$ for all sufficiently large k, and so $x^{-2/n} \le x^{c_k} \le x^{+2/n}$ for all sufficiently large k. Taking $k \to \infty$ then gives $x^{-2/n} \le x^\gamma \le x^{+2/n}$. Consequently, if $\gamma_k \to \gamma$ as $k \to \infty$ with γ_k *real* numbers, for any positive integer n we have $|\gamma_k - \gamma| \le 1/n$ for sufficiently large k. Therefore $x^\gamma x^{-2/n} = x^{\gamma - 2/n} \le x^{\gamma_k} \le x^{\gamma + 2/n} = x^\gamma x^{+2/n}$. The squeeze theorem then shows that $x^\gamma = \lim_{k \to \infty} x^{\gamma_k}$. The case where $0 < x < 1$ can be handled in a similar manner to that used above. Finally,

$$
\begin{aligned}
(x^\alpha)^\beta &= \lim_{k \to \infty} \left(\lim_{\ell \to \infty} x^{a_\ell} \right)^{b_k} = \lim_{k \to \infty} \lim_{\ell \to \infty} (x^{a_\ell})^{b_k} \\
&= \lim_{k \to \infty} \lim_{\ell \to \infty} x^{a_\ell b_k} = \lim_{k \to \infty} x^{\alpha b_k} = x^{\alpha\beta},
\end{aligned}
$$

as we wanted.

$\square$

This proof is fairly long because we had to deal with each step separately: positive whole number to integers, integers to rationals, rationals to reals. We had to use some theorems from analysis in going from integers to rationals (as the value x^a could be irrational even if neither x nor a are irrational), and of course in going from rational to real exponents.

3.10 Exercises

(1) Suppose that $\phi \colon \mathbb{R} \to \mathbb{R}$ is a convex function. Show that
$$
\psi(x) = \sup_y xy - \phi(y)
$$
is also a convex function (although it may have the value $+\infty$).

(2) Show that if $f \colon \mathbb{R} \to \mathbb{R}$ is a differentiable convex function, then for any $x, y \in \mathbb{R}$, $f(x) + f'(x)(y - x) \le f(y)$. The converse holds since for $z = \theta x + (1 - \theta)y$, $f(x) \ge f(z) + f'(z)(x - z) = f(z) + (1 - \theta)f'(z)(x - y)$ and $f(y) \ge f(z) + f'(z)(y - z) = \theta f'(z)(y - x)$. Multiplying the first inequality by θ and the second by $(1 - \theta)$ and adding gives the desired result.

(3) Show that if $f \colon \mathbb{R} \to \mathbb{R}$ is a twice continuously differentiable convex function, and for all $x \in \mathbb{R}$ we have $f''(x) \ge 0$, then f is convex. [**Hint:** For $y > x$, $f'(y) = f'(x) + \int_x^y f''(t)\, dt \ge f'(x)$ so f' is a non-decreasing function. For $z = \theta x + (1 - \theta)y$, $x < y$ and $0 \le \theta \le 1$, show that the tangent line at $(z, f(z))$ lies under the graph of f at x and y.] Show that f convex implies that $f''(x) \ge 0$ for all x by means of Exercise 2 and Taylor series with 2nd order remainder (Theorem 3.15).

(4) *Wilson's theorem.* Show that if p is an odd prime, then $(p - 1)! \equiv -1$ (mod p). [**Hint:** Pair up each factor with its multiplicative inverse. The only problems are the factors that are their own multiplicative inverses.]

(5) Show that if $2^m + 1$ is prime, then m is a power of two. Such primes are called Fermat primes.

(6) Show that if $2^m - 1$ is prime, then m is prime. Such primes are called Mersenne primes.

(7) Can you generalize either of Exercises 5 or 6 to numbers of the form $b^m \pm 1$ with $b > 2$? Is there a variation on $b^m + 1$ or $b^m - 1$ that might work?

(8) Modify the gcd algorithm to compute not only $d = gcd(a, b)$, but also integers s and t where $d = sa + tb$. This is called the *extended gcd algorithm*.

(9) ⚠ Using the convergence of (3.5) for all $s > 1$, show that we *cannot* have $p_k \geq C k^\alpha$ for all k if $C > 0$ and $\alpha > 1$. Conclude that the number of primes $\leq x$ must grow faster than $C x^{1/\alpha}$ for any $\alpha > 1$ and $C > 0$.

(10) Suppose that $\mathcal{P}$ is a set of points and $\mathcal{L}$ a set of straight lines in the plane $\mathbb{R}^2$ with the following property: every point in $\mathcal{P}$ lies on exactly three lines in $\mathcal{L}$, and every line in $\mathcal{L}$ contains exactly two points in $\mathcal{P}$. Show that $3 |\mathcal{P}| = 2 |\mathcal{L}|$. Use this to show that the smallest possible value for $|\mathcal{P}| = 4$, and draw an example of such a configuration. [**Hint:** To show $3 |\mathcal{P}| = 2 |\mathcal{L}|$, count the number of pairs (p, ℓ) where $p \in \mathcal{P}$ is a point on line $\ell \in \mathcal{L}$.]

(11) Prove that (3.11) makes multiplication of equivalence classes modulo n well-defined.

(12) A *field* is an algebraic structure with two operations, addition and multiplication with the usual rules as for rational numbers; in particular, every element other than zero has a multiplicative inverse. Show that the set of equivalence classes $[a]_{\equiv p}$ modulo p, p a prime number, forms a field under the usual operations. This field is denoted $\mathbb{Z}/p\mathbb{Z}$.

You need to check the following axioms: for all a, b and c: $a + b = b + a$, $a + (b + c) = (a + b) + c$, $a + (-a) = 0$, $0 + a = a$, $a \cdot b = b \cdot a$, $a \cdot (b \cdot c) = (a \cdot b) \cdot c$, $0 \cdot a = 0$, $1 \cdot a = a$, $a \cdot (b + c) = a \cdot b + a \cdot c$, and for every $a \neq 0$ there is a b such that $a \cdot b = 1$.

(13) Consider real numbers together with ∞ with the following two operations: $a \oplus b = \min(a, b)$ and $a \otimes b = a + b$. Show that: $a \oplus b = b \oplus a$, $a \otimes b = b \otimes a$, $(a \oplus b) \oplus c = a \oplus (b \oplus c)$, $(a \otimes b) \otimes c = a \otimes (b \otimes c)$, $(a \oplus b) \otimes c = (a \otimes c) \oplus (b \otimes c)$, $\infty \oplus a = a$, $0 \otimes a = a$, $\infty \otimes a = \infty$, $a \otimes (-a) = 0$ for all a, b and $c \in \mathbb{R} \cup \{\infty\}$. This is known as the $(\min, +)$ algebra.

(14) Prove the alternating series test theorem:

Theorem. *If $a_n > 0$ for all n, and a_n, $n = 1, 2, 3, \dots$ is a decreasing* *sequence with* $\lim_{n\to\infty} a_n = 0$, *then* $\sum_{n=1}^{\infty} (-1)^{n-1} a_n$ *converges.*

[**Hint:** Show that if $s_n = \sum_{k=1}^{n} (-1)^{k-1} a_k$, then s_{2n}, $n = 1, 2, 3, \ldots$, is a decreasing sequence and s_{2n+1}, $n = 1, 2, 3, \ldots$ is an increasing sequence.]

(15) Suppose that $g(x)$ and $h(x)$ are continuous real-valued functions of $x \in [a, b]$. Show that if h does not change sign in $[a, b]$ then $\int_a^b g(x) h(x) \, dx = g(c) \int_a^b h(x) \, dx$ for some $c \in [a, b]$. This is the *generalized mean value theorem.*

(16) ⚠ The usual definition of the determinant of a square matrix $\det(A)$ is the recursive definition

$$\det(A) = \sum_{j=1}^{n} (-1)^{j+1} a_{1j} \det(A(1 \mid j)) \qquad (3.13)$$

where $A(i \mid j)$ is the matrix A with row i and column j removed. To terminate the recursion, for a 1×1 matrix, $\det[a_{11}] = a_{11}$.

An alternative definition of $\det(A)$ is

$$\det(A) = \sum_{\sigma} \text{sign}(\sigma) \prod_{i=1}^{n} a_{i,\sigma(i)} \qquad (3.14)$$

where σ ranges over all permutations of $1, 2, 3, \ldots, n$. Note that $\text{sign}(\sigma) = +1$ if σ can be written as a composition of an even number of swaps, and -1 if it can be written as a composition of an odd number of swaps. Show that this is equivalent to (3.13).

(17) A *lattice* is a set S with a partial order $\preceq$ and the following properties: for any two $x, y \in S$ there are $u, v \in S$ where $x, y \preceq u$ and $x, y \preceq r$ implies $u \preceq r$, and also $v \preceq x, y$ and $r \preceq x, y$ implies $r \preceq v$. We say that u is the *join* of x and y (denoted $x \vee y$) and v is the *meet* of x and y (denoted $x \wedge y$). Show that the following are examples of lattices, identifying the join and meet of any given pair of elements of the lattice:

(a) $S = \mathbb{R}$ with the usual order relationship $\leq$.

(b) $S = \mathcal{P}(A)$, the set of subsets of A, with the "subset or equal to" relation $\subseteq$.

(c) S is the set of convex subsets of $\mathbb{R}^n$, with the "subset or equal to" relation $\subseteq$.

(d) $S = \mathbb{R}^2$ with the order relationship $\mathbf{x} \preceq \mathbf{y}$ if and only if $x_i \leq y_i$ for all i.

(18) Exercise 2.22 is about proving that $\forall \epsilon > 0 (|a - b| < \epsilon) \Rightarrow a = b$. Use this to justify why $0.99999\ldots = 1.00000\ldots$.

(19) The *supremum* of a set $S \subset \mathbb{R}$ is the least upper bound L of S: $\forall x \in S \, (x \leq L)$ and if for any number L', $\forall x \in S \, (x \leq L')$ then $L \leq L'$. Show that if S is bounded above $(\exists M \, \forall x \in S \, (x \leq M))$ then S has a supremum by the following process: Pick $a_1 \in S$ and let $b_1 = M$ so that $\exists x \in S \, (a_1 \leq x)$ but $\forall x \in S \, (x \leq b_1)$. For any given positive integer n, put $c_n = (a_n + b_n)/2$. If $\forall x \in S \, (x \leq c_n)$ then set $b_{n+1} = c_n$ and $a_{n+1} = a_n$; otherwise set $a_{n+1} = c_n$ and $b_{n+1} = b_n$. Show that $|b_n - a_n| \leq 2^{1-n} |b_1 - a_1|$ for all n, and that $\exists x \in S \, (a_n \leq x)$ and $\forall x \in S \, (x \leq b_n)$ for all n. Also show that $a_n \leq a_{n+1} \leq b_{n+1} \leq b_n$ for all n. Now you can use the result that *bounded increasing sequences converge* to show that $\lim_{n \to \infty} a_n$ and $\lim_{n \to \infty} b_n$ exist; show that these limits are the same, and call this limit L. Show that L is the desired supremum. Check that there is only one supremum: show that if L_1 and L_2 are two suprema of S, then $L_1 = L_2$. Note that if S is *not* bounded above, we say "$\sup S = +\infty$".

(20) Show that if $\sup S$ is finite, then $L = \sup S$ if and only if $x \leq L$ for all $x \in S$, and for any $\epsilon > 0$ there is an $x \in S$ where $x > L - \epsilon$.

(21) The limsup (or *limit superior*) of a sequence a_n, $n = 1, 2, 3, \ldots$ of real numbers is denoted $\limsup_{n \to \infty} a_n$ and is equal to $\lim_{n \to \infty} \sup \{a_n, a_{n+1}, a_{n+2}, \ldots\}$. Show that $\limsup_{n \to \infty} a_n$ either exists or is "$+\infty$". [**Hint:** Show that $b_n = \sup \{a_n, a_{n+1}, a_{n+2}, \ldots\}$ for $n = 1, 2, 3, \ldots$ is a decreasing sequence.]

(22) Show that if $\lim_{n \to \infty} a_n = L$ (L a real number) then $\limsup_{n \to \infty} a_n = L$.

(23) Show that if $a_n \leq M$ for all n, then $\limsup_{n \to \infty} a_n \leq M$.

(24) ⚠ For the counterexample in Section 3.8.2, show that $\max_{r \geq 0} a_k(r) = \max_{r \geq 0} r^k e^{-r}/k! = k^k/(k! \, e^k)$ and we can use $b_k = k^k/(k! \, e^k)$. Use Stirling's approximation ($m! \sim \sqrt{2\pi m} \, (m/e)^m$ as $m \to \infty$) to show that $b_k \sim 1/\sqrt{2\pi k}$ and therefore that $\sum_{k=0}^{\infty} b_k = +\infty$, and so Theorem 3.20 cannot be applied to the counterexample.

(25) A *group action* of a group G on a set X is a function $\pi \colon G \times X \to X$ where $\pi(g, \pi(h, x)) = \pi(g * h, x)$ and $\pi(e, x) = x$ for all $g, h \in G$ and $x \in X$ where e is the identity of the group G ($e * g = g * e = g$ for all $g \in G$). The *orbit* of $x \in X$ is the set $Gx = \{\pi(g, x) \mid g \in G\}$. Show that $\{Gx \mid x \in X\}$ is a *partition* of X; that is, show that if $Gx \cap Gy \neq \emptyset$ then $Gx = Gy$, and $\bigcup_{x \in X} Gx = X$.

(26) For a group action $\pi \colon G \times X \to X$ let $G_x = \{g \in G \mid \pi(g, x) = x\}$. Show that G_x is a subgroup of G. The subgroup G_x is called the *stabilizer subgroup for $x \in X$*.

(27) For a group action $\pi \colon G \times X \to X$, show that for any $x \in X$ we have $|G| =$

$|Gx|\,|G_x|$. [**Hint:** Start with $|G| = \sum_{y \in Gx} |\{\, g \in G \mid \pi(g,x) = y \,\}|$.] This is known as the *orbit–stabilizer theorem*.

(28) If H is a subgroup of G, we can create a relation $g_1 \sim g_2$ if and only if $g_1 * g_2^{-1} \in H$. Show that this is an equivalence relation on G. If H is a finite group, show that the number of elements of an equivalence class for this equivalence relation is $|H|$, the number of elements of H.

(29) With the equivalence relation from the previous exercise, show that $g_1 \sim k_1$ and $g_2 \sim k_2$ implies $g_1 * g_2 \sim k_1 * k_2$ for all $g_1, g_2, k_1, k_2 \in G$ if and only if $g^{-1} * H * g = H$ for all $g \in G$. (In this case, we say that H is a *normal subgroup* of G.)

(30) Prove that the series $\sum_{n=0}^{\infty} \sum_{k=0}^{n} x^k y^{n-k}/(k!\,(n-k)!)$ converges absolutely so that infinite rearrangements of the sum are justified.

(31) For the triangle shown below, the law of sines is that $a/\sin A = b/\sin B = c/\sin C$. Prove this by computing the area of the triangle in different ways, using different edges as the base and different edges as the hypotenuse of a right-angled triangle for computing the perpendicular height.

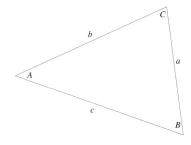

Chapter 4

More advanced proof-making

4.1 Counterexamples and proofs

Counterexamples complement proofs. Counterexamples explicitly show how a statement can be false. They can also guide writing a proof because they point out dead ends and false leads. However, finding counterexamples can be hard. In fact, to create counterexamples, we should look to proven facts to guide us. This means there is a two-way relationship between counterexamples and proofs.

4.1.1 *Minimizing functions of one variable*

A well-known theorem about minimizing a function $f \colon [a,b] \to \mathbb{R}$ is as follows:

Theorem 4.1. *If $f \colon [a,b] \to \mathbb{R}$ is differentiable then a minimizer is either a or b or a point x^* where $f'(x^*) = 0$.*

That a minimizer exists follows from the Bolzano–Weierstrass theorem: since f is differentiable it is also continuous; continuous real-valued functions on closed and bounded sets have minimizers. For more on the usefulness of knowing that a minimizer *exists*, see Section 4.7. The question is, how can we find a minimizer?

Proof. Suppose that x^* minimizes f over the interval $[a,b]$. Suppose $x^* \neq a, b$. Since f is differentiable, then $f'(x^*)$ exists. If $f'(x^*) > 0$ then, set $\epsilon = \frac{1}{2} f'(x^*) > 0$. Then there is a $\delta > 0$ where $|x - x^*| < \delta$ and $a \leq x \leq b$ implies

$$\left| \frac{f(x) - f(x^*)}{x - x^*} - f'(x^*) \right| < \epsilon = \frac{1}{2} f'(x^*),$$

so

$$-\frac{1}{2} f'(x^*) < \frac{f(x) - f(x^*)}{x - x^*} - f'(x^*) < \frac{1}{2} f'(x^*),$$

and

$$\frac{1}{2}f'(x^*) < \frac{f(x) - f(x^*)}{x - x^*} < \frac{3}{2}f'(x^*).$$

Choosing $x = \max(a, x^* - \frac{1}{2}\delta) \in [a, b]$ gives $0 \neq x - x^* < 0$ and so

$$0 < \frac{1}{2}f'(x^*)(x^* - x) < f(x^*) - f(x) < \frac{3}{2}f'(x^*)(x^* - x)$$

and $f(x) < f(x^*)$ contradicting the assumption that x^* minimizes f. Similarly, we can show that $f'(x^*) < 0$ also leads to a contradiction. Thus $x = a$ or $x = b$ or $f'(x^*) = 0$.   □

This result is often referred to as Fermat's theorem, even though Fermat claimed it before calculus was invented! That theorem was not hard, but can we do better? Could we replace it with the following?

Conjecture. *If $f : [a, b] \to \mathbb{R}$ is differentiable then the minimizer is a point x^* where $f'(x^*) = 0$.*

We should not believe this. But can we nail down why it is wrong? If we want to look for a counterexample, let us look for a function where $f'(x) \neq 0$ for any x. One such function is $f(x) = e^x$ or $f(x) = 1/(x + 1)$ on the interval $[0, 1]$. Clearly these functions are continuous on the interval $[0, 1]$, so a minimizer exists (in the first case it is at $x = 0$; in the second case it is at $x = 1$).

What about the following?

Conjecture. *If $f : [a, b] \to \mathbb{R}$ is continuous then either a minimizer is a or b or a point x^* where $f'(x^*) = 0$.*

This time we dropped the *differentiability* assumption. So we should look for a continuous but not (everywhere) differentiable function. Continuous but *nowhere* differentiable functions exist, but they are hard to construct, so we will try to avoid those. But we want differentiability to fail exactly at the minimizer. Looking at the proof of Theorem 4.1, we just need differentiability at the minimizer x^* to get $f'(x^*) = 0$. Can we find such a function? A standard example of a function that it not everywhere differentiable is the absolute value function: $f(x) = |x|$ which is not differentiable at $x = 0$. To be a counterexample, we do not want $x = 0$ to be an endpoint of the interval, so take $[a, b] = [-1, +1]$. The minimizer of $f(x) = |x|$ in $[-1, +1]$ is then at $x = 0$, and the derivative is undefined.

What about this refinement?

Conjecture. *If $f : [a, b] \to \mathbb{R}$ is continuous then either a minimizer is a or b or a point x^* where $f'(x^*) = 0$ or $f'(x^*)$ is undefined.*

None of the counterexamples we have found so far contradict this conjecture. In fact, we now have a theorem: the proof would start *"We prove this by contradiction. Suppose x^* is a minimizer where $a < x^* < b$ and $f'(x^*)$ is defined ..."* The rest of the proof is like the proof of Theorem 4.1.

4.1.2 *Interchanging limits*

Often in a starting course in real analysis it would be nice to use a rule like:

Conjecture. *Provided all limits exist,*

$$\lim_{m \to \infty} \lim_{n \to \infty} a_{mn} = \lim_{n \to \infty} \lim_{m \to \infty} a_{mn}.$$

Unfortunately, it is false. We might try to write a proof of this conjecture, using various strategies, but come up short. But how can we clearly identify this conjecture as false? What kind of double sequence a_{mn} could we use for a counterexample? If we tried $a_{mn} = b_m + c_n$ then (provided all limits exist), $\lim_{m \to \infty} \lim_{n \to \infty} (b_m + c_n) = \lim_{m \to \infty} (b_m + \lim_{n \to \infty} c_n) = (\lim_{m \to \infty} b_m) + (\lim_{n \to \infty} c_n)$ and $\lim_{n \to \infty} \lim_{m \to \infty} (b_m + c_n) = \lim_{n \to \infty} ((\lim_{m \to \infty} b_m) + c_n) = (\lim_{m \to \infty} b_m) + (\lim_{n \to \infty} c_n)$ and the two limits are equal. In the same way, if $a_{mn} = b_m c_n$ we get

$$\lim_{m \to \infty} \lim_{n \to \infty} b_m c_n = (\lim_{m \to \infty} b_m)(\lim_{n \to \infty} c_n) = \lim_{n \to \infty} \lim_{m \to \infty} b_m c_n.$$

So a counterexample will have to "mix-up" m and n. Polynomials will not work since we would just get $\pm \infty$ as the limits (and technically, do not exist). So we need to use division. Here we might need some trial and error. Here is one attempt: $a_{mn} = m/n$. Then $\lim_{m \to \infty} \lim_{n \to \infty} (m/n) = \lim_{m \to \infty} 0$ but $\lim_{n \to \infty} \lim_{m \to \infty} (m/n) = \lim_{n \to \infty} \infty = \infty$, at least formally. So far (formally) we have different limits. But a true counterexample will have to have finite limits. How about $a_{mn} = 1/(1 + (m/n))$? The "$1 +$" in the denominator avoids division by zero. Note that this can also be represented as $a_{mn} = n/(m + n)$. Then $\lim_{m \to \infty} \lim_{n \to \infty} n/(m + n) = \lim_{m \to \infty} 1 = 1$ while $\lim_{n \to \infty} \lim_{m \to \infty} n/(m + n) = \lim_{n \to \infty} 0 = 0$.

Creating good counterexamples usually involves applying the knowledge that you have already gained. There may be trial and error, but the guidance of established theorems is often essential.

Fig. 4.1: Rooms at the Infinity Hotel

4.2 Dealing with the infinite

There is always room at the Infinity Hotel! Whenever someone new arrives asking for a room, the order goes out: the person in room n moves to room $n + 1$ so the new arrival can stay in room one. But wait, an infinite busload of tired travelers just arrived! How can we accommodate them? Make the order: the person in room n must move to room $2n$. That way all the new arrivals can be accommodated in the odd-numbered rooms.

Clearly infinite sets behave differently to finite sets. Most of the time, this kind of bizarre behavior of infinite sets does not cause difficulty. But sometimes it does.

4.2.1 *Rearrangements of conditionally convergent series*

An infinite series $\sum_{n=1}^{\infty} a_n$ is *convergent* if the sequence of partial sums $s_N = \sum_{n=1}^{N} a_n$ converges to a limit as $N \to \infty$. The series $\sum_{n=1}^{\infty} a_n$ is called *absolutely convergent* if $\sum_{n=1}^{\infty} |a_n|$ converges; if $\sum_{n=1}^{\infty} a_n$ is convergent but is not absolutely convergent, we say that it is *conditionally convergent*.

A *rearrangement* of the series $\sum_{n=1}^{\infty} a_n$ is a series $\sum_{k=1}^{\infty} a'_k$ where for each k, $a'_k = a_n$ for exactly one value of n, and for each n, $a'_k = a_n$ for exactly one value of k.[1] A rearrangement is a *finite rearrangement* if $a'_k = a_n$ implies $k = n$ except for a finitely many values of k.

Write $s_N = \sum_{n=1}^{N} a_n$ and $s'_N = \sum_{k=1}^{N} a'_k$ where the a'_k are a rearrangement of the a_n. If the rearrangement is a finite rearrangement, then there is an N_0 where $n \geq N_0$ implies $a'_n = a_n$. Then if $N \geq N_0$ we must have $s_N = s'_N$, and so one series converges if and only if the other converges, and the series $\sum_{n=1}^{\infty} a_n$ and $\sum_{k=1}^{\infty} a'_k$ have exactly the same values.

The difficulties lie with infinite rearrangements.

Theorem 4.2. *If $\sum_{n=1}^{\infty} a_n$ is a conditionally convergent series of real numbers, then for any real number L there is a rearrangement where $\sum_{k=1}^{\infty} a'_k = L$.*

---

[1]This definition assumes that the values a_n are all different. In practice, this is not a serious problem. An alternative definition would be in terms of a one-to-one correspondence $\psi \colon \mathbb{N} \to \mathbb{N}$ and we take $a'_k = a_{\psi(k)}$.

The key to understanding this is to split the terms a_n into positive and negative terms: let b_k be the kth positive number in the sequence a_n, and c_n be the kth number in the sequence with $a_n \leq 0$. Then $\sum_{k=0}^{\infty} b_k = +\infty$ and $\sum_{k=0}^{\infty} c_k = -\infty$. You can check that if one of $\sum_{k=0}^{\infty} b_k$ and $\sum_{k=0}^{\infty} c_k$ were finite, the other must be as well, in which case the original series would be absolutely convergent: a contradiction. If $L > 0$, then we take as many b_k's as needed until $\sum_{n=1}^{N} b_k > L$. This is possible because $\sum_{k=1}^{\infty} b_k = +\infty$. Then choose as many c_k's as needed until $\sum_{k=1}^{N} b_k + \sum_{k=1}^{M} c_k < L$. Then add in more b_k's until we just pass L again: $\sum_{k=1}^{N} b_k + \sum_{k=1}^{M} c_k + \sum_{k=N+1}^{N'} b_k > L$, and then add in more c_k's until we just pass L again: $\sum_{k=1}^{N} b_k + \sum_{k=1}^{M} c_k + \sum_{k=N+1}^{N'} b_k + \sum_{k=M+1}^{M'} c_k < L$. Repeating this process we can alternate between values above L and values below L. From convergence of $s_N = \sum_{n=1}^{N} a_n$, it follows that $a_{N+1} = s_{N+1} - s_N \to 0$ as $N \to \infty$. This means that the distance we go beyond L in each direction goes to zero as we include more and more terms. Thus this new series does indeed converge to L.

To formalize the proof, we need an argument by induction. The hardest part is finding a suitable notation for constructing the rearrangement.

On the other hand, convergent series of positive numbers behave much better.

Theorem 4.3. *If $\sum_{n=1}^{\infty} a_n$ is convergent and $a_n \geq 0$ for all n, then for any rearrangement a_k', we have $\sum_{n=1}^{\infty} a_n = \sum_{k=1}^{\infty} a_k'$.*

Proof. Let $s = \sum_{n=1}^{\infty} a_n$ and $s_N = \sum_{n=1}^{N} a_n$. Also let $s' = \sum_{k=1}^{\infty} a_k'$ and $s_M' = \sum_{k=1}^{M} a_k'$. There is a function $\psi \colon \mathbb{N} \to \mathbb{N}$ that is one-to-one and onto where $a_n = a_k'$ whenever $n = \psi(k)$. Let $\theta(k) = \max\{\psi(j) \mid 1 \leq j \leq k\}$ and $\rho(k) = \min\{\ell \in \mathbb{N} \mid \ell \notin \psi(\{1, 2, \ldots, k\})\}$, so that for any k

$$\{1, 2, \ldots, \rho(k) - 1\} \subseteq \{\psi(1), \psi(2), \ldots, \psi(k)\} \subseteq \{1, 2, \ldots, \theta(k)\}.$$

Note that for a set S, we use the notation $\psi(S) = \{\psi(x) \mid x \in S\}$. Then

$$\sum_{n=1}^{\rho(k)-1} a_n \leq \sum_{j=1}^{k} a_j' = \sum_{j=1}^{k} a_{\psi(j)} \leq \sum_{n=1}^{\theta(k)} a_n,$$

as all terms are ≥ 0. We now need to show that $\rho(k), \theta(k) \to \infty$ as $k \to \infty$.

Note that $\theta(k) \geq k$ as otherwise ψ would not be one-to-one. Thus $\theta(k) \to \infty$ as $k \to \infty$.

This uses that fact that ψ is one-to-one. We also need to use the fact that ψ is onto.

Note that

$$n \notin \psi\left(\{1, 2, \ldots, k\}\right) \implies n \geq \min\left\{\ell \in \mathbb{N} \mid \ell \notin \psi\left(\{1, 2, \ldots, k\}\right)\right\}.$$

Also $\rho(k+1) \geq \rho(k)$ since $\psi\left(\{1, 2, \ldots, k\}\right) \subset \psi\left(\{1, 2, \ldots, k, k+1\}\right)$.

Now $\lim_{k\to\infty} \rho(k) = \infty$. To see this, suppose otherwise: $\rho(k)$ would be a bounded increasing sequence and therefore convergent; let $M = \lim_{k\to\infty} \rho(k)$. Since $\rho(k)$ can only have integer values, M is also an integer and there must be a value K where $k \geq K$ implies $|\rho(k) - M| < 1$, so $\rho(k) = M$. Then $M \notin \psi\left(\{1, 2, \ldots, k\}\right)$ for all $k \geq K$. That is, $M \neq \psi(k)$ for any value of k, which contradicts the assumption that ψ is onto.

Now since $\rho(k) - 1, \theta(k) \to \infty$ as $k \to \infty$, $\lim_{k\to\infty} \sum_{n=1}^{\rho(k)-1} a_n = \sum_{n=1}^{\infty} a_n$ and $\lim_{k\to\infty} \sum_{n=1}^{\theta(k)} a_n = \sum_{n=1}^{\infty} a_n$. By the squeeze theorem, $\lim_{k\to\infty} \sum_{j=1}^{k} a'_j = \sum_{n=1}^{\infty} a_n$ as well, which is what we wanted to prove. $\qquad\square$

The idea of this proof is to use the squeeze theorem so we needed to get upper and lower bounds on the partial sums of the series $\sum_{j=1}^{\infty} a'_j$. To show that the upper and lower bounds have the same limit, we need to use the fact that the function defining the rearrangement is one-to-one and onto. Both of these are needed to make the proof work.

The theorem about rearrangements of convergent positive series is a universal statement about rearrangements: *any* rearrangement will give the same sum. The theorem about conditionally convergent series is an existence theorem: *there is a rearrangement such that* .... Such a proof usually involves a construction: we have to build a rearrangement with a peculiar property. We first need an idea about how to build it, and then do so via induction, or by means of a readily available existence theorem.

4.2.2   *Shift operators*

If V is a *finite-dimensional* vector space and $L\colon V \to V$ is a linear function, then L is onto if and only if L is one-to-one (or, if you like, range$(L) = V$ if and only if $\ker(L) = \{0\}$). But this is not true for infinite dimensional spaces, and the Infinity Hotel gives a clue as to how to do it.

First, let us create an infinite dimensional vector space, which also has a measure of "size" or "distance":

$$\ell^2 = \left\{ \mathbf{x} = (x_1, x_2, x_3, \ldots) \mid \text{each } x_i \in \mathbb{R} \text{ and } \sum_{i=1}^{\infty} x_i^2 \text{ is finite} \right\}.$$

The measure of "size" that we use is appropriate to the definition of the space:

$$\|\mathbf{x}\| = \sqrt{\sum_{i=1}^{\infty} x_i^2}.$$

It can be checked that this satisfies the properties of a *norm*: $\|\mathbf{x}\| \geq 0$ for all $\mathbf{x}$; $\|\mathbf{x}\| = 0$ implies $\mathbf{x} = 0$; $\|\alpha\mathbf{x}\| = |\alpha|\,\|\mathbf{x}\|$ for all $\alpha \in \mathbb{R}$ and all $\mathbf{x}$; and finally, $\|\mathbf{x} + \mathbf{y}\| \leq \|\mathbf{x}\| + \|\mathbf{y}\|$ for all $\mathbf{x}$ and $\mathbf{y}$. The distance between two points $\mathbf{x}$ and $\mathbf{y}$ is $\|\mathbf{y} - \mathbf{x}\|$. We use the notation $\mathbf{e}_1 = (1, 0, 0, \ldots)$, $\mathbf{e}_2 = (0, 1, 0, \ldots)$, $\mathbf{e}_3 = (0, 0, 1, \ldots)$, etc.

Taking a hint from our story of the extra guest at the Infinity Hotel, here is a linear function $\ell^2 \to \ell^2$ that has some new properties not found in linear functions $V \to V$ with V finite dimensional. The guest in room k has to move to room $k + 1$ to make room for the previous guest. In analogy, let us take

$$S\mathbf{e}_k = \mathbf{e}_{k+1}, \qquad k = 1, 2, 3, \ldots.$$

That is, for $\mathbf{x} = (x_1, x_2, x_3, \ldots) = \sum_{k=1}^{\infty} x_k \mathbf{e}_k$, we have

$$S\mathbf{x} = \sum_{k=1}^{\infty} x_k \mathbf{e}_{k+1} = (0, x_1, x_2, x_3, \ldots).$$

Now this linear function $\ell^2 \to \ell^2$ is *not* onto: $S\mathbf{x} \neq (1, 0, 0, 0, \ldots)$ for any $\mathbf{x} \in \ell^2$. On the other hand, $S\mathbf{x} = 0$ would mean that $0 = x_1 = x_2 = x_3 = \cdots$ giving $\mathbf{x} = 0$ so S is one-to-one.

We can reverse the process (the guest in room one of the Infinity Hotel checks out, and the guest in room $k + 1$ moves to room k) and so create a new linear function $T \colon \ell^2 \to \ell^2$:

$$T\mathbf{e}_{k+1} = \mathbf{e}_k, \qquad k = 1, 2, 3, \ldots.$$

But what about $T\mathbf{e}_1$? If the guest in room one checks out, we could use $T\mathbf{e}_1 = 0$. That is,

$$T\mathbf{x} = (x_2, x_3, x_4, \ldots).$$

Then $\ker T$ is generated by $\mathbf{e}_1$, yet T is onto! To see that T is onto, consider the equation $T\mathbf{x} = \mathbf{b}$. This has solutions $\mathbf{x} = (x_1, b_1, b_2, b_3, \ldots)$ for any $x_1 \in \mathbb{R}$.

4.3 Bootstrapping

Sometimes proofs are lengthy because we need to go through a sequence of steps, showing first that a property holds in simpler cases. Once enough simpler cases are put together, the result can be shown in the general case. This is very much like "building bad things from nice things" as described in Section 3.9. In fact, the example of that section (Theorem 3.23, showing that $x^a x^b = x^{a+b}$, $(x^a)^b = x^{ab}$, and $x^{-a} = 1/x^a$ for real a and b) is an example of bootstrapping: first we show the results to be true for positive integer exponents, then for integer exponents, then for rational exponents, and finally for real exponents.

4.3.1   *For square matrices,* $\det(AB) = \det(A)\,\det(B)$

The product formula for determinants of square matrices is very often proven in a step-by-step way where A is a so-called elementary matrix. Elementary matrices are the result of a single elementary row operation applied to the identity matrix

$$I = \begin{bmatrix} 1 & 0 & 0 & \cdots & 0 \\ 0 & 1 & 0 & \cdots & 0 \\ 0 & 0 & 1 & \cdots & 0 \\ \vdots & \vdots & \vdots & \ddots & \vdots \\ 0 & 0 & 0 & \cdots & 1 \end{bmatrix}.$$

There are three different kinds of elementary row operations:

(1) Swapping a pair of rows.
(2) Multiplying a row by a scale factor $\alpha \neq 0$.
(3) Adding a multiple of one row to another row.

It is easy to show that if the same elementary row operation applied to I gives E and applied to B gives B', then $B' = EB$. That is, the result of an elementary row operation applied to B is equal to the product of an elementary matrix E and B. Then for each elementary matrix E we need to compute $\det E$, $\det B$ and $\det B'$ to check that $\det B' = \det E \det B$. Of course, we have to deal with each kind of elementary matrix in a case-by-case approach. To complete the proof, we need to show that A is a product of elementary matrices: $A = E_m E_{m-1} \cdots E_2 E_1$. Then

$$\det(AB) = \det(E_m E_{m-1} \cdots E_2 E_1 B)$$
$$= \det E_m \det E_{m-1} \cdots \det E_2 \det E_1 \det B.$$

On the other hand,

$$\det(A) = \det(E_m E_{m-1} \cdots E_2 E_1)$$
$$= \det E_m \det E_{m-1} \cdots \det E_2 \det E_1.$$

So $\det(AB) = \det(A)\,\det(B)$.

These kinds of proofs are usually fairly long, since we have to treat each kind of elementary matrix separately. Each case is a small proof in itself.

4.4 Impredicative definitions

Impredicative definitions are definitions that give no finite way of determining if a particular thing actually satisfies the definition.

For example, a *positive definite matrix* is a real square $n \times n$ matrix A where $\mathbf{z}^T A \mathbf{z} > 0$ for all $\mathbf{z} \neq 0$ in $\mathbb{R}^n$.[2] Checking that a matrix is positive semi-definite directly from the definition would involve checking uncountably many vectors $\mathbf{z} \in \mathbb{R}^n$. Fortunately there are other tests we can use to see if a matrix is positive definite that can be carried out in finite time.

Another example is the *convex hull* of a set $A \subset \mathbb{R}^n$, which is the smallest convex set containing A. Here "smallest" is with regard to the inclusion ordering "$\subseteq$". An alternative definition of convex hull (which guarantees that the convex hull of A exists and is unique) is that it is the intersection of all convex sets containing A. This alternative definition is also impredicative.

Impredicative definitions can be difficult to work with in some cases, but they give the essential *properties* of the objects concerned. These easily lead to proofs:

Theorem 4.4. *If A and B are positive definite matrices and $\alpha > 0$, then so are αA and $A + B$.*

Proof. Suppose A and B are positive definite matrices, $\alpha > 0$ and $\mathbf{z} \neq 0$.

Then $\mathbf{z}^T (\alpha A) \mathbf{z} = \alpha(\mathbf{z}^T A \mathbf{z}) > 0$ being the product of two positive numbers.

Also, $\mathbf{z}^T (A + B) \mathbf{z} = \mathbf{z}^T A \mathbf{z} + \mathbf{z}^T B \mathbf{z} > 0$ being the sum of two positive numbers.

Since these are true for any $\mathbf{z} \neq 0$, αA and $A + B$ are positive definite. $\square$

We can also show consequences of this property:

Theorem 4.5. *If A is positive definite, then A is invertible.*

Proof. If A is positive definite then, by definition, A is square. Now A is invertible if and only if the only solution of $A\mathbf{x} = 0$ is $\mathbf{x} = 0$.

Suppose $A\mathbf{x} = 0$. Then $\mathbf{x}^T A \mathbf{x} = \mathbf{x}^T 0 = 0 \not> 0$, and since A is positive definite, $\mathbf{x} = 0$. Thus A is invertible. $\square$

There is an equivalent definition for being positive definite that is finitely testable, which is known as *Sylvester's criterion*:

Theorem 4.6. *A real matrix A is positive definite if and only if the matrix $C = \frac{1}{2}(A + A^T)$ satisfies the following conditions:*

---

[2] Some definitions of "positive definite" assume that A is also symmetric ($A^T = A$). We do not do that here.

- $c_{11} > 0$,
- $\det \begin{bmatrix} c_{11} & c_{12} \\ c_{21} & c_{22} \end{bmatrix} > 0$,
- $\det \begin{bmatrix} c_{11} & c_{12} & c_{13} \\ c_{21} & c_{22} & c_{23} \\ c_{31} & c_{32} & c_{33} \end{bmatrix} > 0$,

$$\vdots$$

- $\det C > 0$.

Sylvester's criterion enables us to quickly tell that, for example, $\begin{bmatrix} 7 & -2 \\ -2 & 1 \end{bmatrix}$ is positive definite, but $\begin{bmatrix} 4 & -2 \\ -2 & 1 \end{bmatrix}$ is not. On the other hand, given Sylvester's criterion for a matrix as the definition of positive definite matrices, it would be far from obvious that A and B positive definite implied $A + B$ positive definite. The route connecting the definition of positive definite and Sylvester's criterion is indirect.

Below we prove Sylvester's criterion. We do not launch and all-out attack from the beginning. Rather, let us first explore the idea of positive definite matrices with some lemmas.

Lemma 4.1. *The square matrix A is positive definite if and only if $C = \frac{1}{2}(A + A^T)$ is positive definite.*

Proof. The crucial calculation is this:

$$\mathbf{z}^T A \mathbf{z} = (\mathbf{z}^T A \mathbf{z})^T \quad \text{since } \mathbf{z}^T A \mathbf{z} \text{ is a } 1 \times 1 \text{ matrix,}$$
$$= \mathbf{z}^T A^T \mathbf{z}^{TT} = \mathbf{z}^T A^T \mathbf{z}.$$

Then

$$\mathbf{z}^T C \mathbf{z} = \mathbf{z}^T \frac{1}{2}(A + A^T)\mathbf{z} = \frac{1}{2}\left(\mathbf{z}^T A \mathbf{z} + \mathbf{z}^T A^T \mathbf{z}\right) = \mathbf{z}^T A \mathbf{z}$$

for all $\mathbf{z}$. In particular, for $\mathbf{z} \neq 0$, if C positive definite, then $\mathbf{z}^T A \mathbf{z} = \mathbf{z}^T C \mathbf{z} > 0$ so A positive definite; if A positive definite, then $\mathbf{z}^T C \mathbf{z} = \mathbf{z}^T A \mathbf{z} > 0$ for all $\mathbf{z}$, so C is positive definite. $\qquad \square$

The following lemma shows how we can make a positive definite matrix even more positive definite.

Lemma 4.2. *If A is positive definite and $\alpha \geq 0$, then $A + \alpha I$ is also positive definite.*

Proof. If $\mathbf{z} \neq 0$ then $\mathbf{z}^T(A+\alpha I)\mathbf{z} = \mathbf{z}^T(A\mathbf{z}+\alpha\mathbf{z}) = \mathbf{z}^T A\mathbf{z}+\alpha\mathbf{z}^T\mathbf{z} \geq \mathbf{z}^T A\mathbf{z} > 0$, so $A + \alpha I$ is also positive definite. $\qquad\square$

Now we show that A being positive definite implies that $\det(A) > 0$. This is the last determinant considered in Sylvester's criterion, so $\det(A) > 0$ is a necessary condition for A to be positive definite.

Lemma 4.3. *If A positive definite, then $\det(A) > 0$.*

Proof. By Theorem 4.5, A positive definite implies that it is invertible, so $\det(A) \neq 0$. That is $\det(A) < 0$ or $\det(A) > 0$.

> *How can we show that $\det(A)$ is not negative? We can use a limiting argument for $\det(A + \alpha I)$ as $\alpha \to +\infty$ as we know that $\det(I) = +1$ is positive.*

The determinant function is a continuous function of the matrix entries, so $\det(A + \alpha I)$ is a continuous function of α. But $\det(A + \alpha I) \neq 0$ for any $\alpha \geq 0$ since $A + \alpha I$ is positive definite for $\alpha \geq 0$ by Lemma 4.2. The sign of $\det(A + \alpha I)$ must be the same for all $\alpha \geq 0$; otherwise, by the intermediate value theorem, there would be an $\alpha \geq 0$ where $\det(A + \alpha I) = 0$, which would be a contradiction.

Note that

$$\det(A + \alpha I) = \det(\alpha[\alpha^{-1}A + I]) = \alpha^n \det(\alpha^{-1}A + I).$$

As $\alpha \to +\infty$, $\alpha^{-1}A+I \to I$. Therefore, $\alpha^{-n}\det(A+\alpha I) = \det(\alpha^{-1}A+I) \to \det(I) = +1 > 0$ as $\alpha \to \infty$. So there must be some large $\alpha > 0$ where $\det(A + \alpha I) > 0$. As the sign of $\det(A + \alpha I)$ cannot change for $\alpha \geq 0$, we have $\det(A) > 0$ as we wanted. $\qquad\square$

We can get the intermediate *necessary* conditions for A to be positive definite:

Theorem 4.7. *If A is an $n \times n$ positive definite matrix, then*

$$\det \begin{bmatrix} a_{11} & a_{12} & \cdots & a_{1k} \\ a_{21} & a_{22} & \cdots & a_{2k} \\ \vdots & \vdots & \ddots & \vdots \\ a_{k1} & a_{k2} & \cdots & a_{kk} \end{bmatrix} > 0$$

for $k = 1, 2, \ldots, n$.

Proof. If A is an $n \times n$ positive definite matrix, we can write

$$A = \begin{bmatrix} A_{11} & A_{12} \\ A_{21} & A_{22} \end{bmatrix} \quad \text{where } A_{11} = \begin{bmatrix} a_{11} & a_{12} & \cdots & a_{1k} \\ a_{21} & a_{22} & \cdots & a_{2k} \\ \vdots & \vdots & \ddots & \vdots \\ a_{k1} & a_{k2} & \cdots & a_{kk} \end{bmatrix}.$$

We show that A_{11} is positive definite. Let $\mathbf{z}_1 \in \mathbb{R}^k$ and $\mathbf{z}_1 \neq 0$. Then

$$\mathbf{z}_1^T A_{11} \mathbf{z}_1 = \begin{bmatrix} \mathbf{z}_1 \\ 0 \end{bmatrix}^T \begin{bmatrix} A_{11} & A_{12} \\ A_{21} & A_{22} \end{bmatrix} \begin{bmatrix} \mathbf{z}_1 \\ 0 \end{bmatrix} = \mathbf{z}^T A \mathbf{z} > 0$$

where $\mathbf{z} = [\mathbf{z}_1^T, 0^T]^T \neq 0$. Thus A_{11} is positive definite. By Theorem 4.3, $\det(A_{11}) > 0$ as we wanted.   □

So far in our proofs, we have not used anything about symmetry. This is crucially important for the sufficiency part of Sylvester's criterion: Consider the counterexample

$$A = \begin{bmatrix} 1 & 2 \\ -2 & -1 \end{bmatrix}; \tag{4.1}$$

then $\det[a_{11}] = 1 > 0$ and $\det(A) = 1 \times (-1) - (-2) \times 2 = +3 > 0$ which satisfy Sylvester's conditions except symmetry, but A is not positive definite since

$$\begin{bmatrix} 0 \\ 1 \end{bmatrix}^T \begin{bmatrix} 1 & 2 \\ -2 & -1 \end{bmatrix} \begin{bmatrix} 0 \\ 1 \end{bmatrix} = -1. \tag{4.2}$$

Now we are ready for a proof for Sylvester's criterion. We use induction on the size of the matrix, as follows.

Proof of Theorem 4.6. We prove that Sylvester's criterion implies A is positive definite by induction on n for $n \times n$ matrices. (The necessity of Sylvester's criterion follows from Lemma 4.1 and Theorem 4.7 already proved in this section.)

Base case: $n = 1$: If $n = 1$ then an $n \times n$ matrix A consists of a single number: $A = [a_{11}]$, which is positive definite if and only if $a_{11} = c_{11} > 0$.

Suppose true for $n = k$; **show true for** $n = k + 1$: Suppose that A is a $(k + 1) \times (k + 1)$ matrix, and $C = \frac{1}{2}(A + A^T)$, so that $C = C^T$ and A satisfies Sylvester's criterion. We show that C is positive definite. Write

$$C = \begin{bmatrix} \widehat{C} & \mathbf{c} \\ \mathbf{c}^T & \gamma \end{bmatrix}$$

where $\widehat{C}$ is a $k \times k$ matrix, $\mathbf{c} \in \mathbb{R}^k$ and $\gamma \in \mathbb{R}$. Note that by the induction hypothesis, $\widehat{C}$ is positive definite and $\det \widehat{C} > 0$. $\widehat{C}$ is also symmetric. Sylvester's criterion also implies that $\det C > 0$. Noting that

$$C = \begin{bmatrix} I & 0 \\ \mathbf{c}^T \widehat{C}^{-1} & 1 \end{bmatrix} \begin{bmatrix} \widehat{C} & \mathbf{c} \\ 0 & \gamma - \mathbf{c}^T \widehat{C}^{-1} \mathbf{c} \end{bmatrix},$$

we see that

$$\det C = \det \begin{bmatrix} I & 0 \\ \mathbf{c}^T \widehat{C}^{-1} & 1 \end{bmatrix} \det \begin{bmatrix} \widehat{C} & \mathbf{c} \\ 0 & \gamma - \mathbf{c}^T \widehat{C}^{-1} \mathbf{c} \end{bmatrix}$$

$$= \det(I)\, 1\, (\det \widehat{C})\, (\gamma - \mathbf{c}^T \widehat{C}^{-1} \mathbf{c}) > 0,$$

so $\gamma > \mathbf{c}^T \widehat{C}^{-1} \mathbf{c} \geq 0$. Writing $\mathbf{z} = [\widehat{\mathbf{z}}^T, \zeta]^T$,

$$\mathbf{z}^T C \mathbf{z} = \begin{bmatrix} \widehat{\mathbf{z}} \\ \zeta \end{bmatrix}^T \begin{bmatrix} \widehat{C} & \mathbf{c} \\ \mathbf{c}^T & \gamma \end{bmatrix} \begin{bmatrix} \widehat{\mathbf{z}} \\ \zeta \end{bmatrix}$$

$$= \widehat{\mathbf{z}}^T \widehat{C} \widehat{\mathbf{z}} + \zeta \left(\mathbf{c}^T \widehat{C} \widehat{\mathbf{z}} + \widehat{\mathbf{z}}^T \widehat{C} \mathbf{c} \right) + \gamma \zeta^2$$

$$= (\widehat{\mathbf{z}} + \zeta \mathbf{c})^T \widehat{C} (\widehat{\mathbf{z}} + \zeta \mathbf{c}) - \zeta^2 \mathbf{c}^T \widehat{C} \mathbf{c} + \gamma \zeta^2,$$

using a version of "completing the square". Then

$$\mathbf{z}^T C \mathbf{z} = (\widehat{\mathbf{z}} + \zeta \mathbf{c})^T \widehat{C} (\widehat{\mathbf{z}} + \zeta \mathbf{c}) + (\gamma - \mathbf{c}^T \widehat{C} \mathbf{c}) \zeta^2 \geq 0$$

for any $\mathbf{z}$ as $\widehat{C}$ positive definite and $\gamma - \mathbf{c}^T \widehat{C} \mathbf{c} > 0$. If $\mathbf{z}^T C \mathbf{z} = 0$ we have to have $\widehat{\mathbf{z}} + \zeta \mathbf{c} = 0$ and $\zeta = 0$. But this implies $\widehat{\mathbf{z}} = 0$ and $\zeta = 0$, so that $\mathbf{z} = 0$. Thus for all $\mathbf{z} \neq 0$ we have $\mathbf{z}^T C \mathbf{z} > 0$ and therefore C is positive definite.

Conclusion: For $n = 1, 2, 3, \ldots$ if A is an $n \times n$ matrix that satisfies Sylvester's criterion, then A is positive definite. $\qquad \square$

A natural question is: How could we have come up with Sylvester's criterion from the impredicative definition of *positive definite*? Clearly that would not be easy. In Sylvester's day, determinants were the way to answer questions in linear algebra, and there would be a natural question whether A positive definite implies $\det(A) > 0$ (Lemma 4.3). The converse is not true, for a 2×2 diagonal matrix with -1's on the diagonal has positive determinant but is not positive definite. Looking at submatrices (Theorem 4.7) gives another hint. The counterexample (4.1) also hints about the need for symmetric matrices or something similar. Reducing the problem to the symmetric case is clearly possible from Lemma 4.1.

In sum, then, it would be difficult, but not impossible to come up with Sylvester's criterion through a series of natural questions and partial results. Note that counterexamples are crucial for identifying where conditions need to be added.

4.5   Diagonal proofs

Self-reference is a common feature in certain parts of mathematics. Diagonal proofs make use of self-reference to create objects that have certain properties, usually of a negative kind. These objects are sometimes called *self-defeating objects*.

A first example of an abstract kind is the following result:

Theorem 4.8. *If A is a set, then the set of subsets of A, denoted $\mathcal{P}(A)$, cannot be put in one-to-one correspondence with A. In fact, there is no onto map $A \to \mathcal{P}(A)$.*

Proof. Consider a function $\phi\colon A \to \mathcal{P}(A)$. We will show that this is not onto; that is, there is a set $B \subseteq A$ where $B \neq \phi(a)$ for any $a \in A$.

> *Now we construct a self-defeating object B; something that cannot be in the image of ϕ. The idea is related to Russell's object that cannot be a set (2.17).*

Let $B = \{\, b \in A \mid b \notin \phi(b) \,\}$. Suppose that $B = \phi(a)$ for some a. Then either $a \in B$ or $a \notin B$. If $a \in B$, then by definition of B, $a \notin \phi(a) = B$, which is a contradiction. If $a \notin B = \phi(a)$ then we must have $a \in B$ by definition of B. This is impossible, so $B \neq \phi(a)$ for any $a \in A$ and ϕ is not onto, and so cannot be a one-to-one correspondence. $\square$

> Note that the very definition of B ensures that it cannot be in the range of ϕ. This result has a number of consequences. For example,

Theorem 4.9. $\mathbb{R}$ *is uncountable.*

Proof. Take $A = \mathbb{N} = \{1, 2, 3, \ldots\}$, the natural numbers. Saying that $\mathbb{R}$ is uncountable is equivalent to saying that there is no one-to-one correspondence $\psi\colon \mathbb{N} \to \mathbb{R}$.

Suppose there is a one-to-one correspondence $\psi\colon \mathbb{N} \to \mathbb{R}$. We can create an onto map $\theta\colon \mathbb{R} \to \mathcal{P}(\mathbb{N})$ by

$$\theta(x) = \{\, k \mid k \geq 1 \text{ and } d_k = 5 \,\},$$

where $x = (d_{-m}d_{-m+1}\cdots d_0.d_1 d_2 \cdots)_{10}$ is the decimal expansion of x. (Note that this function is well-defined in spite of the usual ambiguity of $0.999999\cdots = 1.000000\ldots$.) To see that θ is onto, note that a set $E \subseteq \mathbb{N}$ is equal to $\theta(x_E)$ where

$$x_E = 5 \times \sum_{k \in E} 10^{-k}.$$

Now $\widehat{\psi} := \theta \circ \psi \colon \mathbb{N} \to \mathcal{P}(\mathbb{N})$ would be onto as it is the composition of two onto functions. But this is impossible by Theorem 4.8. $\qquad\square$

4.5.1 *Russell paradox*

An especially famous example of a "self-defeating object" of the kind considered here is due to the logician and philosopher, Bertrand Russell. The story behind Russell's example starts with the work of Gottlob Frege, a German mathematician and logician who essentially invented predicate calculus as described in Chapter 2. Frege described a new system for logic in his book *Begriffsschrift* (*Concept-script*) (Frege, 1879), and developed the idea in a two-volume sequence *Grundgesetze der Arithmetik* (*Basic Laws of Arithmetic*). Just when he was ready to send the second volume to the printer in 1903, he received a letter from Russell with Russell's example. The example showed that the system he had built contained a contradiction, and could not be used as a foundation for mathematics (or logic). Ways around this paradox were found by various logicians, including Bertrand Russell who used this insight to create a new foundation for mathematics with Alfred Whitehead published as the three-volume set *Principia Mathematica* (Whitehead and Russell, 1925–1927).

So what is Russell's example? Sets are commonly constructed as the collection of elements satisfying a certain condition: $S = \{\, x \mid P(x) \,\}$ where $P(x)$ is a predicate. But, unless we can show that S is contained in an already constructed set, this might not give a valid set. In particular, consider the Russell "set" (2.17):

$$R = \{\, x \mid x \notin x \,\}.$$

That is, R is the "set" of all things that are not elements of themselves. Since most objects are not elements of themselves, R should be a big "set".

Unfortunately we have a paradox here: Suppose $R \in R$. Then by definition of R, $R \notin R$ which contradicts our assumption. So $R \notin R$. But then by definition of R, we R must contain R; that is, $R \in R$, which is also impossible!

The way out of this paradox is to declare that to ensure that when we define a set, we do not allow an unrestricted formula to define a set. Instead we need some set U that is guaranteed to contain all the items we are interested in, and then we can define a set S by

$$S = \{\, x \in U \mid P(x) \,\}$$

for a predicate P. This avoids the kind of negative self-reference that is the heart of Russell's example. Russell's later work with Whitehead in *Principia Mathematica* also avoids this problem by creating a hierarchy of predicates (predicates of elements, predicates of predicates of elements, predicates of predicates of predicates of elements, etc.). This system, known as the *theory of types*, is not much used today but was an important milestone in the development of mathematical logic.

4.5.2 *Gödel's incompleteness theorem*

The logician Kurt Gödel is best remembered for his incompleteness theorem that says, roughly, that any consistent logical system powerful enough to include Peano's axioms for the natural numbers must have a statement that can neither be proven or disproven in that system. The essential idea is to first encode formulas and proofs as positive integers, and then develop (or show existence of) formulas that represent when a statement is provable. From there it is a short step to creating a self-defeating formula that essentially says that it is not provable.

We need a technical assumption called ω-*consistency*: for any true existential statement $\exists z\, P(z)$ there is a positive whole number n where $P(n)$ is true. This is a stronger condition than consistency as models of the arithmetic of natural numbers can include infinite "integers" as well as the usual ones.

Here is an outline of how his proof works: letters $(a, b, c, \ldots, x, y, z, A, B, C, \ldots)$, symbols $(+, -, \times, (,), \neg, \wedge, \vee, \Rightarrow, \forall, \exists, \ldots)$ and digits $(0, 1, 2, \ldots, 9)$ are encoded as natural numbers. A formula is a sequence of these letters, symbols, and digits which we can represent as a list of natural numbers $c_1, c_2, \ldots, c_N$. This list can be represented as a single number $p_1^{c_1} p_2^{c_2} \cdots p_N^{c_N}$ where p_k is the kth prime number. For a given formula ϕ, we represent its numerical encoding by $\lceil \phi \rceil$.

Formulas for predicates can be developed using this encoding, such as Formula(x) which is true if x is the encoding of a syntactically correct formula, Axiom(x) which is true if x is the encoding of an axiom, and Infer(x, y, z) if z is the encoding of a formula that can be inferred from formulas with encodings x and y. We can extend this idea to proofs: proofs are lists of formulas where each formula is either an axiom or can be inferred from previous formulas in the proof, perhaps with "annotations" describing the inferences. If the formulas (and annotations) in the proof are encoded as $f_1, f_2, \ldots, f_M$, then the entire proof can be encoded as $p_1^{f_1} p_2^{f_2} \cdots p_M^{f_M}$. Then we can create a formula Proof(x, y) which is true if x is the encoding of a proof of a formula with encoding y.

The formula Provable$(y) = \exists x\, \text{Proof}(x, y)$ is true if there is a proof of the formula with encoding y. There is one more formula we need to create: SelfRef(x, z) is true if x is the encoding of a formula $\phi(y)$ with one free variable (call it y), and z is the encoding of the formula $\phi(\lceil \phi(y) \rceil)$. The predicate SelfRef can also be represented as a formula in predicate calculus, with whole number arithmetic, although it would be extremely complex. Now we can set up the self-defeating object: let

$$\text{Godel}(x) = \forall z \; [\text{SelfRef}(x, z) \Rightarrow \neg\text{Provable}(z)],$$
$$G = \text{Godel}(\lceil\text{Godel}(x)\rceil).$$

The statement G is our self-defeating object. Suppose we could prove G. Then Provable($\lceil G \rceil$) would be true. But proving G means proving Godel($\lceil\text{Godel}(x)\rceil$); that is,

$$\forall z \; [\text{SelfRef}(\lceil\text{Godel}(x)\rceil, z) \Rightarrow \neg\text{Provable}(z)].$$

Now SelfRef($\lceil\text{Godel}(x)\rceil, z$) is true if $z = \lceil\text{Godel}(\lceil\text{Godel}(x)\rceil)\rceil = \lceil G \rceil$, so we can conclude that $\neg$Provable($\lceil G \rceil$) and so G is not provable.

Now suppose that we could prove $\neg G$. That is, suppose we could prove

$$\neg\forall z \; [\text{SelfRef}(\lceil\text{Godel}(x)\rceil, z) \Rightarrow \neg\text{Provable}(z)].$$

Noting that $p \Rightarrow q$ is equivalent to $\neg p \lor q$, we could then prove

$$\exists z \; [\text{SelfRef}(\lceil\text{Godel}(x)\rceil, z) \land \neg\neg\text{Provable}(z)].$$

Assuming that the system is ω-consistent there is an actual number $z \in \mathbb{N}$ where SelfRef($\lceil\text{Godel}(x)\rceil, z$) $\land \neg\neg$Provable(z).

But SelfRef($\lceil\text{Godel}(x)\rceil, z$) for $z \in \mathbb{N}$ means that $z = \lceil\text{Godel}(\lceil\text{Godel}(x)\rceil)\rceil = \lceil G \rceil$. Thus Provable($\lceil G \rceil$), and again using ω-consistency, there is an actual number $y \in \mathbb{N}$ where Proof(y, z), and there is a corresponding proof of G.

That is, if we can prove G then we can prove $\neg G$; if we can prove $\neg G$ then we can prove G. As long as our theory is consistent, it is not possible to prove $G \land \neg G$, so neither G nor $\neg G$ can be proven. That is, there are some statements without free variables where neither the statement nor its negation are provable.

While this is not a complete proof, it gives the essence of the Gödel's proof of his first incompleteness theorem. Rosser was able to replace the assumption of ω-consistency with simple consistency. Gödel's second incompleteness theorem takes this further to show that it is not possible to prove a statement that affirms the consistency of natural number arithmetic from within natural number arithmetic.

4.6 Using duality

The idea of duality — pairs of objects that are in some way "dual" to each other — is a powerful idea. The most important property of duals is that the dual of the dual is the original object, or very closely related to it. Duality is often first met in matrix transposes, but there are also dual planar graphs, dual convex functions, dual cones, dual vector spaces, and so on. For a planar graph, the faces of the original graph become the nodes of the dual graph, and edges join nodes in the dual graph if the corresponding faces in the original share a common border or edge. For a real vector space V, the dual vector space V^* is the collection of linear functions $V \to \mathbb{R}$.

4.6.1 *Duality in linear algebra*

A common duality in linear algebra comes from transposes. Below we prove a well-known theorem (Theorem 4.12) about the following vector spaces:

$$\text{range}(A) = \{ \, A\mathbf{x} \mid \mathbf{x} \text{ is a vector} \, \},$$
$$\ker(A) = \{ \, \mathbf{x} \mid A\mathbf{x} = 0 \, \}.$$

For a vector subspace V of $\mathbb{R}^n$, we define

$$V^{\perp} = \left\{ \, \mathbf{x} \in \mathbb{R}^n \mid \mathbf{x}^T \mathbf{y} = 0 \text{ for all } \mathbf{y} \in V \, \right\},$$

which is called the *orthogonal complement* of V.

We need a basic result about dimensions of orthogonal complements:

Theorem 4.10. *If V is a subspace of $\mathbb{R}^n$ then $\dim V^{\perp} = n - \dim V$.*

Theorems of this kind usually involve some kind of messy calculations with bases. Here we take a less direct path, using the idea that linear functions $\mathbb{R}^n \to \mathbb{R}$ also form a vector space of dimension n that is a little less messy.

Proof. Let $\{\mathbf{v}_1, \mathbf{v}_2, \ldots, \mathbf{v}_m\}$ be a basis for V. We can extend this to a basis $\{\mathbf{v}_1, \mathbf{v}_2, \ldots, \mathbf{v}_m, \mathbf{w}_1, \mathbf{w}_2, \ldots, \mathbf{w}_{n-m}\}$ for $\mathbb{R}^n$.

Every vector $\mathbf{z} \in \mathbb{R}^n$ defines a linear function $\mathbb{R}^n \to \mathbb{R}$ given by $\phi(\mathbf{x}) = \mathbf{z}^T \mathbf{x}$. In fact, this gives all the linear functions $\mathbb{R}^n \to \mathbb{R}$: suppose $\phi \colon \mathbb{R}^n \to \mathbb{R}$ is linear. Let $z_i = \phi(\mathbf{e}_i)$ where $\mathbf{e}_i$ is the ith standard basis vector ($\mathbf{e}_1 = [1, 0, 0, \ldots, 0]^T$, $\mathbf{e}_2 = [0, 1, 0, \ldots, 0]^T$, etc.). Then with $\mathbf{z} = [z_1, z_2, \ldots, z_n]^T$ we have $\phi(\mathbf{e}_i) = z_i = \mathbf{z}^T \mathbf{e}_i$. Since ϕ is linear,

$$\phi(\mathbf{x}) = \phi(x_1 \mathbf{e}_1 + x_2 \mathbf{e}_2 + \cdots + x_n \mathbf{e}_n)$$
$$= x_1 \phi(\mathbf{e}_1) + x_2 \phi(\mathbf{e}_2) + \cdots + x_n \phi(\mathbf{e}_n)$$
$$= x_1 z_1 + x_2 z_2 + \cdots + x_n z_n = \mathbf{z}^T \mathbf{x}$$

for any $\mathbf{x} \in \mathbb{R}^n$. Also, if we picked a different vector $\mathbf{u} \neq \mathbf{z}$, then the function $\psi(\mathbf{x}) = \mathbf{u}^T \mathbf{x}$ will be different from the function $\phi(\mathbf{x}) = \mathbf{z}^T \mathbf{x}$. (If you want to do this properly, $\mathbf{u} \neq \mathbf{z}$ means that for some i, $u_i \neq z_i$, so $\psi(\mathbf{e}_i) = u_i \neq z_i = \phi(\mathbf{e}_i)$ and therefore $\psi \neq \phi$.)

Now $\phi(\mathbf{x}) = \mathbf{z}^T\mathbf{x} = 0$ for all $\mathbf{x} \in V$ means that $\mathbf{z} \in V^\perp$. So to understand $V^\perp$ we should look at linear functions that are zero on V.

Now since $\{\mathbf{v}_1, \mathbf{v}_2, \ldots, \mathbf{v}_m, \mathbf{w}_1, \mathbf{w}_2, \ldots, \mathbf{w}_{n-m}\}$ is a basis for $\mathbb{R}^n$, every linear function $\psi\colon \mathbb{R}^n \to \mathbb{R}$ can be represented uniquely in terms of the values $\psi(\mathbf{v}_i)$ and $\psi(\mathbf{w}_j)$: every $\mathbf{x} \in \mathbb{R}^n$ can be written uniquely as $\mathbf{x} = \sum_{i=1}^m c_i \mathbf{v}_i + \sum_{j=1}^{n-m} d_j \mathbf{w}_j$, and

$$\psi\left(\sum_{i=1}^m c_i \mathbf{v}_i + \sum_{j=1}^{n-m} d_j \mathbf{w}_j\right) = \sum_{i=1}^m c_i \psi(\mathbf{v}_i) + \sum_{j=1}^{n-m} d_j \psi(\mathbf{w}_j).$$

We are interested in the linear functions ψ where $\psi(\mathbf{v}) = 0$ for every $\mathbf{v} \in V$. That is, we are interested in the linear functions ψ where $\psi(\mathbf{v}_1) = \psi(\mathbf{v}_2) = \cdots = \psi(\mathbf{v}_m) = 0$. These functions have the form

$$\psi\left(\sum_{i=1}^m c_i \mathbf{v}_i + \sum_{j=1}^{n-m} d_j \mathbf{w}_j\right) = \sum_{j=1}^{n-m} d_j \psi(\mathbf{w}_j).$$

So the linear functions ψ where $\psi(\mathbf{v}) = 0$ for all $\mathbf{v} \in V$ are completely determined by the $n - m$ values $\psi(\mathbf{w}_j)$. Furthermore, since the $\mathbf{w}_j$, $j = 1, 2, \ldots, n-m$ are linearly independent, the values $\psi(\mathbf{w}_j)$ can be assigned independent values. Thus the set of linear functions ψ where $\psi(\mathbf{v}) = 0$ for all $\mathbf{v} \in V$ is a subspace of dimension $n - m$. Since the map $\psi \mapsto \mathbf{z}$ is a one-to-one and onto linear function to the vectors $\mathbf{z} \in V^\perp$, $V^\perp$ is a subspace of dimension $n - m$. $\square$

We first show that the orthogonal complement of the orthogonal complement of a subspace of $\mathbb{R}^n$ is the original subspace:

Theorem 4.11. *If V is a subspace of $\mathbb{R}^n$, then $V^{\perp\perp} = V$.*

Proof. First we show that $V \subseteq V^{\perp\perp}$. Suppose that $\mathbf{v} \in V$. If $\mathbf{y} \in V^\perp$, then by definition of $V^\perp$ we have $\mathbf{y}^T\mathbf{x} = 0$ for all $\mathbf{x} \in V$. Since $\mathbf{v} \in V$, we have $\mathbf{v}^T\mathbf{y} = 0$ for all $\mathbf{y} \in V^\perp$. That is, $\mathbf{v} \in (V^\perp)^\perp = V^{\perp\perp}$. Thus $V \subseteq V^{\perp\perp}$.

On the other hand, $\dim V^{\perp\perp} = \dim(V^\perp)^\perp = n - \dim V^\perp = n - (n - \dim V) = \dim V$ by Theorem 4.10.

If there were some $\mathbf{x} \in V^{\perp\perp}$ but $\mathbf{x} \notin V$, then we could add $\mathbf{x}$ to any basis for V to show that $\dim V^{\perp\perp} \geq \dim V + 1$, contradicting $\dim V^{\perp\perp} = \dim V$.

Therefore, $V^{\perp\perp} \subseteq V$.

Combined with $V \subseteq V^{\perp\perp}$ we get $V = V^{\perp\perp}$, as we wanted. $\square$

We can use this to prove an important result about matrices:

Theorem 4.12. *If A is an $m \times n$ matrix with real entries, then*

$$\ker(A) = \mathrm{range}(A^T)^\perp, \quad \text{and}$$
$$\mathrm{range}(A) = \ker(A^T)^\perp.$$

Proof. We prove the first statement: $\ker(A) = \mathrm{range}(A^T)^\perp$ for any matrix A. This is mostly an exercise in unwrapping definitions:

$$\ker(A) = \{\, \mathbf{x} \mid A\mathbf{x} = 0 \,\}$$
$$= \{\, \mathbf{x} \mid \mathbf{y}^T A \mathbf{x} = 0 \text{ for all } \mathbf{y} \,\}$$
$$= \{\, \mathbf{x} \mid \mathbf{x}^T A^T \mathbf{y} = 0 \text{ for all } \mathbf{y} \,\}$$
$$= \{\, \mathbf{x} \mid \mathbf{x}^T \mathbf{z} = 0 \text{ for all } \mathbf{z} \in \mathrm{range}(A^T) \,\}$$
$$= \mathrm{range}(A^T)^\perp.$$

Now for the second statement: $\mathrm{range}(B) = \ker(B^T)^\perp$ for any matrix B. Put $A = B^T$. Then $\ker(B^T) = \mathrm{range}(B^{TT})^\perp = \mathrm{range}(B)^\perp$. Taking the orthogonal complement of both sides gives (using Theorem 4.11) $\ker(B^T)^\perp = \mathrm{range}(B)^{\perp\perp} = \mathrm{range}(B)$, as we wanted. $\qquad\square$

Theorem 4.11 as stated holds for subspaces of $\mathbb{R}^n$, which is how most people would use this kind of result. But we can replace $\mathbb{R}$ by any *field* (see Exercise 4.5 at the end of this chapter), which can be the complex numbers $\mathbb{C}$, the rational numbers $\mathbb{Q}$, or even integers modulo a prime p ($\mathbb{Z}/p\mathbb{Z}$). Since Theorem 4.12 only uses the fact that A has real entries by referring to Theorem 4.11 (which in turn refers to Theorem 4.10). But neither Theorem 4.12 nor Theorem 4.10 actually uses the fact that the vectors are in $\mathbb{R}^n$, just that we can perform all the usual operations on vectors. So these theorems apply to matrices with entries in any other field as well as matrices with real entries.

4.7 Optimizing

There are two ways in which optimization appears in proofs: one is to use optimization methods or techniques to prove some result, such as an inequality. Properties such as convexity are especially useful in this context.

Another way in which optimization appears in proofs is to prove that some object is optimal in some way. Consider the problem of minimizing $f(x)$ over all $x \in S$ where S is a given set and f a given function. That is, we want to find x where

$$x \in S \land \forall y \in S \, (f(x) \le f(y)); \qquad (4.3)$$

we write the condition (4.3) as Optimal(x). Usually we start working on these problems using necessary conditions: Optimal(x) $\Rightarrow$ Nconditions(x). If we want to prove that some conditions are necessary, we commonly work through a contrapositive (see Exercise 2.1):

$$x \in S \wedge \neg\text{Nconditions}(x) \Rightarrow \exists y \in S \left(f(y) < f(x) \right). \tag{4.4}$$

To prove this, if the necessary conditions for $x \in S$ do *not* hold, then we need to find some y where $f(y) < f(x)$. Usually y will be x with a small or simple modification.

Necessary conditions usually narrow the search. If you can find all the objects that satisfy the necessary conditions, then you can usually check the function values for each of these objects and choose the one that gives the smallest value. But you need one more condition: *that an optimizer exists.*

If there is *only one* x that satisfies the necessary conditions, *and an optimizer exists*, then that x must be optimal:

$$(\exists y \, \text{Optimal}(y) \wedge \exists! z \, \text{Nconditions}(z)) \Rightarrow (\text{Nconditions}(x) \Rightarrow \text{Optimal}(x)).$$

The existence requirement $\exists y \, \text{Optimal}(y)$ is not superfluous. Consider the problem of minimizing $f(x) = e^{-x}$ over $x \geq 0$. The usual necessary conditions are that $x = 0$, or $f'(x) = 0$ and $x \geq 0$. But $f'(x) = 0$ is impossible, so we have to pick $x = 0$, which would give the minimum value as $f(0) = e^{-0} = 1$. But this is not the minimum; in fact it is the maximum! For any $x > 0$, $f(x) < f(0) = 1$.

In this section we have two examples. The first example (Hölder inequalities) is an example of using optimization to prove an inequality. The second example (Huffman codes) involves proving that an object generated by an algorithm is optimal.

4.7.1 *Hölder inequalities*

The aim is to prove that if $p, q > 1$ with $1/p + 1/q = 1$, we have the Hölder inequality:

$$\int_a^b f(x)\,g(x)\,dx \leq \left[\int_a^b |f(x)|^p \, dx \right]^{1/p} \left[\int_a^b |g(x)|^q \, dx \right]^{1/q}. \tag{4.5}$$

This will require several steps, starting with Young's inequality (Theorem 3.3): If $a, b \geq 0$, $p, q > 1$, and $1/p + 1/q = 1$, then

$$ab \leq \frac{a^p}{p} + \frac{b^q}{q}. \tag{4.6}$$

This first step might seem a very long way from the result we want. But we have a start that involves pth and qth powers related to a product. And it can be turned into the result we want.

Theorem 4.13. *For any continuous functions* f, g, *and* p, $q > 1$ *where* $1/p + 1/q = 1$,

$$\int_a^b f(x)\, g(x)\, dx \leq \left[\int_a^b |f(x)|^p\, dx\right]^{1/p} \left[\int_a^b |g(x)|^q\, dx\right]^{1/q}. \tag{4.7}$$

Proof. We reduce this to the case of non-negative f and g:

$$\int_a^b f(x)\, g(x)\, dx \leq \left|\int_a^b f(x)\, g(x)\, dx\right|$$

$$\leq \int_a^b |f(x)|\, |g(x)|\, dx.$$

Now we can use our Lemma.

But $|f(x)|\, |g(x)| \leq |f(x)|^p /p + |g(x)|^q /q$, so

$$\int_a^b f(x)\, g(x)\, dx \leq \int_a^b [|f(x)|^p /p + |g(x)|^q /q]\, dx$$

$$= \frac{1}{p} \int_a^b |f(x)|^p\, dx + \frac{1}{q} \int_a^b |g(x)|^q\, dx.$$

At this point things look stuck. But we can introduce a free parameter $\alpha > 0$ *and replace* $f(x)$ *by* $\alpha f(x)$ *and* $g(x)$ *by* $\alpha^{-1}g(x)$.

Replacing $f(x)$ by $\alpha f(x)$ and $g(x)$ by $\alpha^{-1}g(x)$ with $\alpha > 0$ gives

$$\int_a^b f(x)\, g(x)\, dx = \int_a^b \alpha f(x)\, \alpha^{-1}g(x)\, dx$$

$$\leq \frac{1}{p} \int_a^b |\alpha f(x)|^p\, dx + \frac{1}{q} \int_a^b |\alpha^{-1}g(x)|^q\, dx$$

$$= \frac{\alpha^p}{p} \int_a^b |f(x)|^p\, dx + \frac{\alpha^{-q}}{q} \int_a^b |g(x)|^q\, dx. \tag{4.8}$$

We still do not have the result we want, but we can minimize the right-hand side.

Differentiating the bound on the right-hand side with respect to α, in order to minimize the bound (4.8), we find the minimizing α satisfies

$$\alpha^{p-1} \int_a^b |f(x)|^p\, dx - \alpha^{-q-1} \int_a^b |g(x)|^q\, dx = 0.$$

We set

$$\alpha = \left[\frac{\int_a^b |g(x)|^q \, dx}{\int_a^b |f(x)|^p \, dx} \right]^{1/(p+q)}.$$

We do not need to prove that this is the actual minimizer of the bound, we can just use it.

Substituting this value of α into (4.8) gives

$$\int_a^b f(x) \, g(x) \, dx \leq \frac{1}{p} \left[\frac{\int_a^b |g(x)|^q \, dx}{\int_a^b |f(x)|^p \, dx} \right]^{p/(p+q)} \int_a^b |f(x)|^p \, dx$$

$$+ \frac{1}{q} \left[\frac{\int_a^b |g(x)|^q \, dx}{\int_a^b |f(x)|^p \, dx} \right]^{-q/(p+q)} \int_a^b |g(x)|^q \, dx$$

$$= \frac{1}{p} \left[\int_a^b |g(x)|^q \, dx \right]^{p/(p+q)} \left[\int_a^b |f(x)|^p \, dx \right]^{q/(p+q)}$$

$$+ \frac{1}{q} \left[\int_a^b |g(x)|^q \, dx \right]^{p/(p+q)} \left[\int_a^b |f(x)|^p \, dx \right]^{q/(p+q)}$$

$$= \left[\int_a^b |g(x)|^q \, dx \right]^{p/(p+q)} \left[\int_a^b |f(x)|^p \, dx \right]^{q/(p+q)}.$$

But if we multiply $1/p + 1/q = 1$ by pq we get $q + p = pq$. Then we note that $p/(p+q) = p/(pq) = 1/q$ and $q/(p+q) = q/(pq) = 1/p$. So

$$\int_a^b f(x) \, g(x) \, dx \leq \left[\int_a^b |g(x)|^q \, dx \right]^{1/q} \left[\int_a^b |f(x)|^p \, dx \right]^{1/p}$$

as we wanted. $\square$

Note that

$$\psi(\alpha) := \frac{\alpha^p}{p} \int_a^b |f(x)|^p \, dx + \frac{\alpha^{-q}}{q} \int_a^b |g(x)|^q \, dx$$

is a convex function of $\alpha > 0$, so the necessary condition $\psi'(\alpha) = 0$ is also a sufficient condition. The proof does not use this fact, but its truth partly explains why the proof succeeds.

4.7.2 *Huffman coding*

If we wish to transmit an English letter ('*a*', '*b*', '*c*', ..., '*z*') we can use no more than $\lceil \log_2(26) \rceil = 5$ bits, a bit being a zero or one. We can do better for the *average* length of the string of bits used, but this depends on the probabilities for the different letters: '*e*' is much more common than '*x*', so by using a shorter bit string for '*e*' and a longer bit string for '*x*' we can reduce the average length of the bit strings used to represent the letters. We assume that each letter has a positive probability.

The question is: *How should we encode the letters to minimize the average length of the encoding?*

Huffman's algorithm provides the solution by creating a binary tree. A tree is a connected undirected graph that has no cycle. For definitions related to graphs, see page 15. A rooted tree is a tree with a special node called the root r. For each node v other than the root r there is a unique simple path (without repetitions of nodes or edges) from v to r. The node adjacent to v on this path is the *parent* of v in the tree. If u is a parent of v then we say that v is a *child* of u. Note that each node other than the root has a unique parent, but a parent node can have several children. (See Exercise 4.17.) A node in a rooted tree is called a *leaf* if it has no children. All leaves in a tree have degree one. (See Exercise 4.18.) A *binary tree* is a rooted tree where every node has degree ≤ 3; that is, each node has no more than two children. An *ancestor* of a node x is a node y where the simple path from x to the root contains y. For a given node x of a tree, there is a subtree $T(x)$ which has the nodes $V(x) := \{x\} \cup \{ y \mid x \text{ is an ancestor of } y \}$ and the edges of T between nodes in $V(x)$; we make x the root of $T(x)$.

An example of a binary tree is shown in Figure 4.2. In this tree, r is the root, which has children a and g. The leaf nodes are c, d, f, and j. The subtree $T(b)$ has the nodes b, c, d and edges connecting b and c, and connecting b and d.

How can a binary tree give a code for the letters? First, the leaves of the tree are labeled with the letters. Each edge is labeled either '0' or '1' so that if a node is the parent of two (child) vertices, one of the edges to a child node is labeled with '0' and the other is labeled with '1'. The code representing a letter can be found by reading off the labels of the edges of the simple path from the root to the leaf for that letter. To decode a binary string, we start at the root of the tree and for each bit from left to right, we take the edge labeled by the bit in the string. When we come to a leaf, we have decoded the letter for the binary string so far.

For example, for the tree in Figure 4.2, the leaf node f is represented by the binary code 011, corresponding to the path r, a, e, f from the root r to f.

To minimize the average length of the encoding of the letters, we want to

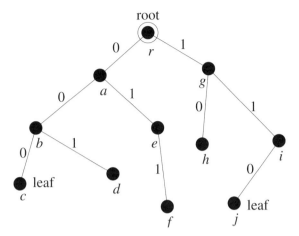

Fig. 4.2: Example of binary tree

minimize the average length from the root of the tree to the leaves, where the probability of the each leaf is the probability of the corresponding letter. The objective function to minimize is therefore

$$\phi(T, \mathbf{p}) = \sum_{x \text{ leaf of } T} p_x \, (\text{distance from } x \text{ to root in } T), \qquad (4.9)$$

where p_x is the probability of the letter associated with the leaf x. Only the leaves of the tree T have probabilities.

To create an optimal tree, Huffman's algorithm works from the bottom up: we start with the leaves (or letters) and at each step we join together the trees with the smallest total probabilities: let $\pi(T, \mathbf{p})$ be the sum of the probabilities of all leaves of a tree T:

$$\pi(T, \mathbf{p}) = \sum_{x \text{ leaf of } T} p_x.$$

Huffman(**p**) /* **pre:** p_i is probability of letter i */
    for $i \leftarrow 1, 2, \ldots,$ length(**p**)
      $T_i \leftarrow$ one vertex tree labelled by i
    $\mathcal{Q} \leftarrow \{\, T_i \mid i = 1, 2, \ldots,$ length(**p**) $\}$
    while $|\mathcal{Q}| > 1$
      /* **pre:** every $T \in \mathcal{Q}$ defines an optimal code
        for the leaves of T */
      let R, S be trees in $\mathcal{Q}$ with smallest $\pi(R, \mathbf{p}), \pi(S, \mathbf{p})$
      $\mathcal{Q} \leftarrow \mathcal{Q} \backslash \{R, S\}$

```
            root
          /   \
  Y ← R      S        /*  π(Y, p) = π(R, p) + π(S, p)  */
  Q ← Q ∪ {Y}
end while
/*  now |Q| = 1 so Q has only one element */
let Q = {T}
return T
```

In trying to prove that the output from this algorithm is indeed optimal, we need to compare the resulting tree with *any* binary tree. It is hard to say anything useful about *every* binary tree beyond the definition.

We need some information about what optimal trees look like. Here is a basic necessary condition for optimal trees.

Lemma 4.4. *If T is a binary tree representing an optimal coding, then every node in T that is not a leaf has exactly two children.*

Proof. We use proof by contradiction. Suppose there is some node u of T that has at least one child node but not two nodes. Since T is a binary tree, u must have exactly one child node; call it v. We can create a new tree T' with the same nodes as T except u, and the edges of T' are the same as for T except that $T(u)$ is replaced by $T(v)$. The root of T' is the root of T, which is denoted r. If x is a leaf of T but not $T(u)$ then the distance from x to r in T' is equal to the distance from x to r in T. If x is a leaf of $T(u)$ then the distance from x to r in T' is one less than the distance from x to r in T. Thus $\phi(T', \mathbf{p}) = \phi(T, \mathbf{p}) - \sum_{x \text{ leaf of } T(u)} p_x < \phi(T, \mathbf{p})$. Thus T is not optimal.

That is, if T is optimal, then every node of T that is not a leaf has exactly two children.   □

Thus, *if there is an optimal tree*, every node in it that is not a leaf must have exactly two children. But can we be sure that there is an optimal tree? The binary trees we are considering have the same number of leaves as the number of elements of our set of letters. Unfortunately, there are infinitely many of these as we can have a long chain of edges going to a leaf:

root •   •   • leaf. But we can use the argument in Lemma 4.4 to reduce the set of binary trees we consider to binary trees where every node that is not a leaf has two children.

Inside Lemma 4.4 we have the following argument: if T is a binary tree where there is a node with only one child, then by replacing that node with its child, we have a new binary tree T' with $\phi(T', \mathbf{p}) < \phi(T, \mathbf{p})$. We do not guarantee that T' is a binary tree where every node is either a leaf or has two children. But if T' has a node with one child, we can repeat the process to give a new tree T'' with $\phi(T'', \mathbf{p}) < \phi(T', \mathbf{p}) < \phi(T, \mathbf{p})$, and so on. The danger here is that this process can go on forever without reaching a tree $T^{(m)}$ where all its nodes are either leaves or have two children. Fortunately we can rule this out since with each of these steps, the number of nodes of the tree is reduced by one. Thus we cannot carry out more steps of this procedure than there are nodes in T; the process must stop, and we have a binary tree $T^{(m)}$ where all nodes are either leaves or have two children, and $\phi(T^{(m)}, \mathbf{p}) < \phi(T, \mathbf{p})$.

Let $\mathcal{B}_n$ denote the set of binary trees with n leaves identified as letters, and $\mathcal{B}_{F,n}$ the subset of $\mathcal{B}_n$ where each node is either a leaf or has two children. (The subscript "F" was chosen for "full".) Then for any tree $T \in \mathcal{B}_n$, there is a tree $T_F \in \mathcal{B}_{F,n}$ with $\phi(T_F, \mathbf{p}) < \phi(T, \mathbf{p})$. Thus we only need to look in the set $\mathcal{B}_{F,n}$ for the optimal tree. Note that while the number of nodes of a tree $T \in \mathcal{B}_n$ is unbounded, there is a definite bound on the number of nodes of a tree $T \in \mathcal{B}_{F,n}$. This means that as there are essentially a finite number of trees in $\mathcal{B}_{F,n}$, there must be an optimal tree.

On the other hand, there is no "worst" tree: given any binary tree T, by inserting nodes in edges, we can always create a new tree T' with $\phi(T', \mathbf{p}) > \phi(T, \mathbf{p})$.

With this in hand, we can go on to the main theorem. As a notational convenience we will use $d(v, w; T)$ to represent the distance between nodes v and w in the tree T.

Theorem 4.14. *The tree generated by the Huffman algorithm generates an optimal coding.*

Proof. The proof is by induction on the number of letters in the alphabet, which is the number of leaves in the tree.

> *We have an algorithm that builds up the tree step-by-step, so naturally we look for a proof by induction. However, we end up doing induction in a different way from the algorithm.*

The number of binary trees with n leaves is finite, so there must be an optimal coding and an optimal tree.

Base case: $n = 1$: If there is only one leaf then the optimal tree is just a single node, which is the root.

Suppose true for $n = m$; show true for $n = m + 1$: Suppose T is the tree generated by the Huffman algorithm for $m + 1$ letters forming the set L, which correspond to leaves of T. Let R and S be the trees selected from $\mathcal{T}$ in the first loop through the algorithm. Then R contains only the node u and S contains only the node v where $p_x \leq p_y \leq p_z$ for all letters $z \neq x, y$. The resulting tree $Y = x \overset{w}{\wedge} y$ consists of a new root node (denoted w) whose children are nodes x and y.

The remaining loops of the `while` loop in Huffman's algorithm are equivalent to applying Huffman's algorithm to the set of letters $L' = (L \setminus \{x, y\}) \cup \{w\}$, with probabilities $p'_w = p_x + p_y$ and $p'_z = p_z$ for all $z \neq w$ in L'. By the induction hypothesis, the resulting tree T' is optimal for L' with probabilities $\mathbf{p}'$. The tree T computed for the set L is the tree T' with the leaf w replaced by the root of the tree Y given above.

Here we use the fact that there exists an optimal tree.

Suppose T_2 is optimal, so $\phi(T_2, \mathbf{p}) \leq \phi(T, \mathbf{p})$.

We need to relate the structure of T_2 to the structure of T. We do this by creating a new tree T_3 that is at least as good as T_2, but has some of the structure of T.

By Lemma 4.4, every node of T_2 is either a leaf node, or has two children. Let r be the root of T_2. Pick a leaf a of T_2 with the largest distance from r. (There must be a furthest leaf node since there is only a finite number of leaves.) Since there are $m + 1 > 1$ leaf nodes, a must have a parent node u. Since u is not a leaf node, it must have two children (one of which is a). Let b be the other child node. Note that $d(a, r; T_2) = d(u, r; T_2) + 1 = d(b, r; T_2)$. Assuming without loss of generality that $p_a \leq p_b$, this means that $p_x \leq p_a$ and $p_y \leq p_b$. Swapping nodes x with a and y with b in T_2 creates a new binary tree T_3. However, $\phi(T_3, \mathbf{p}) = \phi(T_2, \mathbf{p}) - (d(a, r; T_2) - d(x, r; T_2))(p_a - p_x) - (d(b, r; T_2) - d(y, r; T_2))(p_b - p_y) \leq \phi(T_2, \mathbf{p})$. Since T_2 is optimal, $\phi(T_2, \mathbf{p}) \leq \phi(T_3, \mathbf{p})$, and therefore, $\phi(T_3, \mathbf{p}) = \phi(T_2, \mathbf{p}) \leq \phi(T, \mathbf{p})$.

Now we have to find a modification T'_3 of T_3 to compare with T'.

Now u is the parent of leaves x and y in T_3; w is the parent of leaves x and y in T. Recall that T' is T with the nodes x and y, and their edges to w, removed. Let T'_3 be T_3 with the nodes x and y, and their edges to u removed. Identify w in T with u in T_3. Also recall that by the induction hypothesis, $\phi(T', \mathbf{p}') \leq \phi(T'_3, \mathbf{p}')$. But $\phi(T', \mathbf{p}') = \phi(T, \mathbf{p}) - p_x - p_y \geq \phi(T_3, \mathbf{p}) - p_x - p_y = \phi(T'_3, \mathbf{p}')$. Thus $\phi(T', \mathbf{p}') = \phi(T'_3, \mathbf{p}')$, and so $\phi(T, \mathbf{p}) = \phi(T_3, \mathbf{p}) = \phi(T_2, \mathbf{p})$, showing that T is indeed optimal. $\qquad\square$

It is worth stopping for a moment to review how this proof works. We need to combine information about binary trees with the (very basic) necessary conditions of Lemma 4.4, and the steps of the algorithm. As with most proofs involving algorithms, the proof uses induction. To show that the result is optimal it is helpful (but not necessary) to show that an optimal tree exists.

4.8 Generating functions

Generating functions are a way of packaging a lot of information into a function. Usually a sequence a_0, a_1, a_2, a_3, ... is represented by the function

$$f(x) = a_0 + a_1 x + a_2 x^2 + a_3 x^3 + \cdots = \sum_{k=0}^{\infty} a_k x^k.$$

Often asymptotic information about the coefficients a_k can be obtained from the behavior of the function $f(x)$. Note that it is important that $|a_k| \leq C r^k$ for some $C, r \geq 0$ so that the power series for $f(x)$ converges for some non-zero values of x.

4.8.1 *Recurrences and generating functions*

The Fibonacci numbers are a well known sequence of integers

$$1, \ 1, \ 2, \ 3, \ 5, \ 8, \ 13, \ 21, \ 34, \ 55, \ 89, \ \ldots$$

where the nth Fibonacci number is the sum of the previous two Fibonacci numbers:

$$F_n = F_{n-1} + F_{n-2} \qquad \text{for } n \geq 2 \tag{4.10}$$

with the "starting values" $F_0 = 0$, $F_1 = 1$. The equation (4.10) is a recurrence that we can solve.

We can bound Fibonacci numbers as follows:

Lemma 4.5. *For all $n \geq 0$, $0 \leq F_n \leq 2^n$.*

The proof of this is left as Exercise 4.12.

Theorem 4.15. *The nth Fibonacci number is*

$$F_n = \frac{1}{\sqrt{5}} \left[\left(\frac{1 + \sqrt{5}}{2} \right)^n - \left(\frac{1 - \sqrt{5}}{2} \right)^n \right].$$

This is a well-known formula for the Fibonacci numbers. Here we will see a proof of this using generating functions.

Proof. Let $f(x) = \sum_{n=0}^{\infty} F_n x^n$. By Lemma 4.5, this power series is convergent for $|x| < 1/2$. The recurrence (4.10) means that

$$f(x) = F_0 + F_1 x + \sum_{n=2}^{\infty} F_n x^n$$

$$= x + \sum_{n=2}^{\infty} (F_{n-1} + F_{n-2}) x^n$$

$$= x + x \sum_{n=2}^{\infty} F_{n-1} x^{n-1} + x^2 \sum_{n=2}^{\infty} F_{n-2} x^{n-2}$$

$$= x + x \sum_{k=1}^{\infty} F_k x^k + x^2 \sum_{\ell=0}^{\infty} F_\ell x^\ell \qquad (k = n-1, \, \ell = n-2)$$

$$= x + x \left(f(x) - 0 \right) + x^2 f(x) = 1 - (x + x^2) f(x),$$

using the fact that $F_0 = 0$. Solving for $f(x)$ gives

$$f(x) = \frac{x}{1 - x - x^2}.$$

Now $x^2 + x - 1 = (x - \varphi_1)(x - \varphi_2)$ where $\varphi_1 = -(1 + \sqrt{5})/2$ and $\varphi_2 = (\sqrt{5} - 1)/2$. Using partial fractions

$$\frac{x}{1 - x - x^2} = \frac{A}{1 - x/\varphi_1} + \frac{B}{1 - x/\varphi_2}$$

where A and B satisfy the linear equations $A + B = 0$ and $A/\varphi_2 + B/\varphi_1 = 0$. Solving the equations using $B = -A$ gives $A = 1/\sqrt{5}$ so $B = -A = -1/\sqrt{5}$. Now

$$\frac{1}{1 - x/a} = \sum_{k=0}^{\infty} \frac{x^k}{a^k} \qquad \text{for } |x| < |a|,$$

so

$$f(x) = \frac{1}{\sqrt{5}} \sum_{k=0}^{\infty} \left[\frac{1}{\varphi_2^k} - \frac{1}{\varphi_1^k} \right] x^k \qquad \text{for } |x| < \varphi_2.$$

Note that $\varphi_1 \varphi_2 = -1$ so $1/\varphi_1 = -\varphi_2$ and $1/\varphi_2 = -\varphi_1$. Then

$$f(x) = \frac{1}{\sqrt{5}} \sum_{k=0}^{\infty} \left[\left(\frac{1 + \sqrt{5}}{2} \right)^k - \left(\frac{1 - \sqrt{5}}{2} \right)^k \right] x^k.$$

Identifying the coefficients of x^k with F_k gives the desired result. $\qquad \square$

4.8.2 Catalan numbers

The nth Catalan number C_n is the number of ways of parenthesizing a product of n numbers. For example, when $n = 3$, the product abc can be parenthesized two ways so $C_3 = 2$: $(ab)c$ and $a(bc)$. Also $C_4 = 5$, since we can parenthesize $abcd$ as $(ab)(cd)$, $a((bc)d)$, $a(b(cd))$, $((ab)c)d$, or $(a(bc))d$.

To investigate Catalan numbers, we consider the generating function $\mathcal{C}(x) = \sum_{n=1}^{\infty} C_n x^n$. We will be able to obtain a nice expression for $\mathcal{C}(x)$, and from it deduce a formula for C_n. But to do that, we need a recurrence for C_n.

Lemma 4.6. *For $n \geq 2$,*

$$C_n = \sum_{k=1}^{n-1} C_k\, C_{n-k}.$$

Also, $C_1 = 1$.

Proof. Suppose we have a parenthesization of $a_1 a_2 \cdots a_n$. The outer-most parentheses must have the form $(a_1 a_2 \cdots a_k)(a_{k+1} \cdots a_n)$ for some k. Since there are C_k ways of parenthesizing $a_1 a_2 \cdots a_k$ and C_{n-k} ways of parenthesizing $a_{k+1} \cdots a_n$ and k can have any value from one through $n - 1$, we have

$$C_n = \sum_{k=1}^{n-1} C_k\, C_{n-k}.$$

Finally, simple inspection shows that $C_1 = 1$. $\qquad\square$

The sum $\sum_{k=1}^{n-1} C_k\, C_{n-k}$ is closely related to the function $\mathcal{C}(x)^2$:

$$\mathcal{C}(x)^2 = \left(\sum_{k=1}^{\infty} C_k x^k \right) \left(\sum_{\ell=1}^{\infty} C_\ell x^\ell \right)$$

$$= \sum_{k=1}^{\infty} \sum_{\ell=1}^{\infty} C_k C_\ell\, x^{k+\ell}.$$

Adding the coefficients of x^n, we sum over k and ℓ where $k + \ell = n$. That is, $\ell = n - k$. This gives

$$\mathcal{C}(x)^2 = \sum_{n=2}^{\infty} \left(\sum_{k=1}^{n-1} C_k C_{n-k} \right) x^n = \sum_{n=2}^{\infty} C_n x^n.$$

Note that the first non-zero coefficient for $\mathcal{C}(x)^2$ is for x^2, not x. So

$$\sum_{n=2}^{\infty} C_n x^n = \mathcal{C}(x)^2.$$

Therefore,

$$\mathcal{C}(x) = C_1 x + \mathcal{C}(x)^2 = x + \mathcal{C}(x)^2.$$

This is quadratic equation for $\mathcal{C}(x)$. Solving gives

$$\mathcal{C}(x) = \frac{1 \pm \sqrt{1 - 4x}}{2}.$$

Noting that $\mathcal{C}(x) = 0$ and that $\mathcal{C}(x)$ is a continuous function of x for $x < 1/4$, we see that we should choose "$-$" instead of "$+$" in "$\pm$". That is,

$$\mathcal{C}(x) = \frac{1 - \sqrt{1 - 4x}}{2} \qquad \text{for } |x| < \frac{1}{4}.$$

From this we can obtain an explicit formula for C_n using

$$(1 + z)^{\alpha} = 1 + \alpha z + \frac{\alpha(\alpha - 1)}{2!} z^2 + \frac{\alpha(\alpha - 1)(\alpha - 2)}{3!} z^3 + \cdots \qquad \text{for } |z| < 1.$$

For $\alpha = 1/2$,

$$(1 + z)^{1/2} = 1 + \frac{1}{2} z - \frac{1 \cdot 1}{2^2 \, 2!} z^2 + \frac{1 \cdot 1 \cdot 3}{2^3 \, 3!} z^3 - \frac{1 \cdot 1 \cdot 3 \cdot 5}{2^4 \, 4!} z^4 + \cdots$$

$$= 1 + \sum_{k=1}^{\infty} (-1)^{k-1} \frac{(2k - 1)!!}{2^k \, k!} z^k.$$

Then

$$(1 - 4x)^{1/2} = 1 - \sum_{k=1}^{\infty} \frac{(2k - 1)!!}{2^k \, k!} 4^k \, x^k.$$

For more on using generating functions to count combinatorial objects, see *Analytical Combinatorics* (Flajolet and Sedgewick, 2009).

For counting objects where the number of objects a_n of size n grows too fast for $\sum_{n=0}^{\infty} a_n z^n$ to be convergent for any $z \neq 0$ (such as $a_n = n!$ or $a_n = n^n$), we can use exponential generating functions: $g(z) = \sum_{n=0}^{\infty} a_n z^n / n!$.

Note that not only can exact formulas be obtained, but it is often fairly easy to obtain estimates for the a_n, although this often needs complex analysis.

4.9   Exercises

(1) Show that there is a one-to-one and onto mapping $f \colon \mathbb{N} \to \mathbb{N} \times \mathbb{N}$. Also show that the rational numbers $\mathbb{Q}$ are countable (that is, there is an onto mapping $\mathbb{N} \to \mathbb{Q}$). [**Hint:** For $f \colon \mathbb{N} \to \mathbb{N} \times \mathbb{N}$, list the elements (i, j) of $\mathbb{N} \times \mathbb{N}$ along diagonals $i + j = k$ with $k = 2, 3, 4, \ldots$.]

(2) Here is a paradox related to Russell's paradox in a disguised way. A game is a *simple game* if it involves exactly two players who take turns making moves, the results of the moves are completely determined by the rules of the game, and no matter what moves are made by either player, the game terminates after a finite number of moves. Many games are like this. We will introduce a game called "*Super*": the first player chooses a simple game, and then the players play the chosen game, with the second player making the first play in the chosen game.

Theorem. *Super is a simple game.*

Proof. Once a simple game is chosen, there can only be a finite number of moves before the game stops. □

Unfortunately, when Fred and George play *Super*, when Fred goes first he chooses *Super* (he points to the above theorem as evidence that it is indeed a simple game); so George has to choose a simple game, and he also chooses *Super* as well. Continuing in this way, the game *Super* goes on forever. So *Super* is not a simple game!

Explain the source of the paradox in terms similar to Russell's paradox: there is a collection of objects that is too big to be a set.

(3) Show that if $\sum_{n=1}^{\infty} a_n$ is a conditionally convergent series, then for any real number L there is a re-arrangement a_n' of a_n where $\sum_{n=1}^{\infty} a_n' = L$.

(4) Here is one approach to the more advanced theory of Lebesgue integration: We can use Riemann integration to define $\int_a^b f(x)\,dx$ for continuous f. The idea is then to extend the integral to as many functions as reasonably possible. Read "Rethinking the Lebesgue integral" (Lax, 2009).

(5) A *field* is a set with the usual arithmetic operations: $+$ (plus), $-$ (minus), $\times$ (times), and $/$ (divide) with the usual arithmetic properties. (See Exercise 3.12.) Let K denote a field. Examples of fields include the real numbers $\mathbb{R}$, the rationals $\mathbb{Q}$, the complex numbers $\mathbb{C}$, and finite fields such as $\mathbb{Z}/p\mathbb{Z}$ where p is a prime number. A *vector space V over K* is a set of vectors including a zero vector which can be added, and multiplied by elements of K, with the following properties: for all $x,\,y,\,z \in V$ and $\alpha,\,\beta \in K$,

$$
\begin{aligned}
x + (y + z) &= (x + y) + z, & (\alpha\beta)x &= \alpha(\beta x), \\
x + y &= y + x, & (\alpha + \beta)x &= \alpha x + \beta x, \\
\alpha(x + y) &= \alpha x + \alpha y, & 0x &= 0, \\
x + 0 &= x, & 1x &= x.
\end{aligned}
$$

A *linear map* between two vector spaces V and W over K is a function $L: V \to W$ where $L(\alpha x + \beta y) = \alpha L(x) + \beta L(y)$ for any $\alpha\, \beta \in K$ and $x, y \in V$. The *dual space* V' is the set of all linear maps $V \to K$. Show that V' is a vector space over K.

(6) Show that the set of numbers of the form $a + b\sqrt{2}$ with a, b rational numbers forms a field.

(7) Theorem 4.11 applies to finite dimensional vector spaces. Here is a counterexample for infinite dimensional spaces. Consider the vector space $\ell^2 = \{ (x_1, x_2, x_3, \dots) \mid \sum_{k=1}^{\infty} x_i^2 \text{ is finite} \}$ and the vector subspace $V = \{ (x_1, x_2, x_3, \dots) \mid \text{only finitely many } x_i \neq 0 \}$. Check that V really is a vector space, and that $V \subset \ell^2$ but $V \neq \ell^2$. On the other hand, show that $V^\perp = \{0\}$, so that $V^{\perp\perp} = \ell^2 \neq V$.

(8) Suppose that $\{v_1, v_2, \dots, v_m\}$ is a basis for the vector space V, and let $\lambda_i: V \to K$ $(1 \leq i \leq m)$ be given by the formula

$$\lambda_i(c_1 v_1 + c_2 v_2 + \cdots + c_m v_m) = c_i.$$

Show that $\{\lambda_1, \lambda_2, \dots, \lambda_m\}$ is a basis for V'. From this, show that if V is a finite dimensional vector space, then $\dim V = \dim V'$.

(9) Following Exercise 5, show that if $L: V \to W$ is a linear map, then the function $L^*: W' \to V'$ between the dual spaces defined by $L^*(\omega)(v) = \omega(L(v))$ (equivalently, $L^*(\omega) = \omega \circ L$) is a linear map. Note that ω is a linear map $W \to K$ and $L^*(\omega)$ is a function $V \to K$. You need to first show that $L^*(\omega) \in V'$ by showing that $v \mapsto \omega(L(v))$ is a linear map. The map L^* is called the *adjoint* of L.

(10) If V is a vector space over a field K, then there is a function $\mathrm{nat}: V \to V''$ from V to the dual of its dual given by $\mathrm{nat}(x)(\xi) = \xi(x)$ where $\xi \in V'$ is a linear map $V \to K$. (Note that $\mathrm{nat}(x) \in V''$ so $\mathrm{nat}(x)$ is a linear map $V' \to K$.) Show that nat is linear and one-to-one. Show that if V is finite dimensional then nat is onto. The function nat is called the *natural map* $V \to V''$.

(11) If W is a vector subspace of a finite dimensional vector space V over a field K, we let $W^\perp = \{ \phi \in V' \mid \forall w \in W: \phi(w) = 0 \} \subseteq V'$. This is not the orthogonal complement in $\mathbb{R}^n$ since $W^\perp$ is in a different space to W, but should work in the same way. Show that $\dim W^\perp = \dim V - \dim W$. (This proof should look something like Theorem 4.10.) Next show that $\mathrm{nat}(W) \subseteq W^{\perp\perp} \subseteq V''$. Then use the dimensions to show that $\mathrm{nat}(W) = W^{\perp\perp}$.

(12) Prove that for all integers $n \geq 0$, $0 \leq F_n \leq 2^n$. Use this to show that $F(x) = \sum_{n=0}^{\infty} F_n x^n$ is well-defined, at least for $|x| < 1/2$.

(13) Prove Theorem 4.2 using the strategy explained in the text.

(14) Euler's formula for connected planar graphs is $|V| - |E| + |F| = 1$ where V is the set of vertices or nodes, E is the set of edges, and F is the set of faces. Prove this by induction on the number of edges. But be careful how you remove edges! If you remove the only edge connecting a node to the rest of the graph, you need to remove the node as well. You also need to be careful about disconnecting the graph into larger pieces by removing an edge. The only time removing any edge disconnects a graph is when the graph has no cycles, in which case there must be a node of degree one (see Theorem 1.6). Draw some diagrams to think about the problem and how to write the proof. [**Note:** For a finite set S, $|S|$ denotes the number of elements of S.]

(15) Find a counterexample to the conjecture that the product of two real symmetric positive definite matrices is positive definite. [**Hint:** There are 2×2 counterexamples. You can also use the fact that for any symmetric matrix A there is an orthogonal matrix Q where $Q^T A Q$ is diagonal.]

(16) A *coloring* of a graph G is an assignment of colors to the nodes of G where for any edge e in G, the endpoints of e are assigned different colors. If no more than n colors are used, it is called an n-*coloring*. The minimum n for which a graph G has an n-coloring is called the *chromatic number* of G and is denoted $\chi(G)$. Show that the chromatic number of G must be no more than the maximum of $1 + \deg_G(x)$ over all nodes x of G. [**Hint:** Prove this by induction on the number of nodes of G.] Find a counterexample for equality between $\chi(G)$ and $1 + \max_{x \in V(G)} \deg_G(x)$.

(17) Prove that for every vertex u adjacent to a vertex v in a tree with a root r, either u is the parent of v or u is a child of v (that is, v is the parent of u). Furthermore, show that there can only be at most one parent of a vertex v.

(18) Prove that every leaf of a tree has degree one.

(19) A *derangement* is a permutation $\pi: \{1, 2, 3, \ldots, n\} \to \{1, 2, 3, \ldots, n\}$ where $\pi(k) \neq k$ for all k. Show that the number of derangements of n items is given by the recurrence $d_{n+1} = n(d_n + d_{n-1})$ with $d_1 = 0$ and $d_2 = 1$. Find a formula for $D(x) = \sum_{n=1}^{\infty} d_n x^n / n!$.

Chapter 5

Building theories

A large part of the work of many mathematicians is building theories. A theory usually consists of a collection of connected definitions and theorems that lead toward understanding some body of mathematical knowledge. Creating a theory involves crafting definitions as well as identifying theorems to prove. Often this goal is to abstract some well-known mathematical theory beyond its original setting. Sometimes, it comes out of exploring a certain set of ideas.

Here were are not thinking about proving a specific theorem. It is more a matter of deciding what theorems to prove. This is an open-ended activity, and there is often no natural stopping point. Often building a theory is part of a research program: a set of ideas that suggests certain objectives for research.

5.1 Choosing definitions

There is no such thing as an incorrect definition for a new concept. But some definitions are more useful than others. Take, for example, the following definition of *prime number*, which was used in Section 3.2 (Definition 3.3):

Definition. *A positive integer p is a* prime number *if $p \neq 1$ and the only divisors of p are one and p.*

Most of the definition is easy to understand: "*the only divisors of p are one and p*". But why "$p \neq 1$"? This makes one a special case, and may require special attention in proofs. In general, it is undesirable to single out special cases in definitions. But in this case there is an important reason to leave it out: if one were a prime number, then the unique factorization theorem would be false: $1 = 1 \times 1 = 1 \times 1 \times 1$ are different factorizations of one.

Sometimes there are several choices of some basic concepts. For example, graphs or networks can be simple or hyper-graphs, directed or undirected. Should loops (edges which begin and end at the same node) be allowed? Which of the four possible combinations you use depends on the situation. You might start thinking that you should just work with undirected graphs. But if you want to work with flows even in undirected networks, then you will need to assign a direction to each edge.

Suppose we wanted to generalize the idea of "convex set" to the surface of a sphere. There are no "straight lines" on the surface of a sphere, but there are shortest paths. But between opposite points there are infinitely many shortest paths. Should "convex sets" in this context include opposite points? If so, should we require that there is *at least one* shortest path that stays inside the set, or should we require that *all* shortest paths stay inside the set? (If you choose the latter case, then as soon as opposite points are in the set, then the "convex set" must be the entire sphere.)

5.2   What am I modeling?

"But wait!" you say. "I do *pure* mathematics." Yes, you may. But even pure mathematics is directed to understanding some phenomenon or phenomena. For example, the theory of computability is based on some kind of understanding of what *computation* is. The ideas used to describe this concept can be apparently quite different: such as Turing machines, recursive functions, the λ-calculus of Alonzo Church, and Markov algorithms. But they all aim to describe the kinds of things we do when we *compute*. In fact, all these concepts are equivalent.

Galois theory in abstract algebra came out of the question *"Why can't we find a formula for the solution of 5th order polynomials?"* Banach and Hilbert spaces were attempts to find a common approach to understanding differential equations, by making a function a point in a certain (large) space. Set theory is about what kinds of mathematical objects can or should exist (in some sense). Optimization is about very practical tasks of doing the best with what we have. And differential equations came out of models of the real world.

The advantage of thinking about mathematics as a kind of modeling is that it directs your attention to objectives: What are we trying to understand? What kinds of systems are we studying?

With objectives in mind, we can start to formulate a theory. Definitions come from the objects to be studied. After exploring the relationships between them, we can start to formulate conjectures. Then we can work on proving these conjectures, and turn them into theorems.

5.2.1 *Generalizing known concepts*

Many mathematical theories are attempts to generalize what has been proven about simpler objects in more general settings. For example, linear algebra began as the study of systems of linear equations. The central objects of study are matrices (rectangular arrays of numbers) and vectors in $\mathbb{R}^n$ (n numbers stacked vertically). In such a concrete setting, it is easy to do calculations like solving a particular linear system, but hard to formulate and prove relationships. This very concrete way of thinking about linear systems has its limits. Representing the operations performed on a linear system by means of matrix multiplication gives a new way of thinking about these operations.

At first this may seem to be merely a notational change. But even a notational change opens the way for a more abstract way of thinking. Going from

$$\begin{bmatrix} a_{11} & a_{12} & \cdots & a_{1n} \\ a_{21} & a_{22} & \cdots & a_{2n} \\ \vdots & \vdots & \ddots & \vdots \\ a_{n1} & a_{n2} & \cdots & a_{nn} \end{bmatrix} \begin{bmatrix} b_{11} & b_{12} & \cdots & b_{1n} \\ b_{21} & b_{22} & \cdots & b_{2n} \\ \vdots & \vdots & \ddots & \vdots \\ b_{n1} & b_{n2} & \cdots & b_{nn} \end{bmatrix} = \begin{bmatrix} c_{11} & c_{12} & \cdots & c_{1n} \\ c_{21} & c_{22} & \cdots & c_{2n} \\ \vdots & \vdots & \ddots & \vdots \\ c_{n1} & c_{n2} & \cdots & c_{nn} \end{bmatrix}$$

$$c_{ij} = \sum_{k=1}^{n} a_{ik} b_{kj},$$

to

$$AB = C$$

makes it easier to get to $(I + A)^{-1} = I - A + A^2 - A^3 + \cdots$. This is much like the advantage in using algebraic notation $x^2 + 2x + 1 = (x + 1)^2$ over pure English: "the square of the unknown plus twice the unknown plus one is equal to the square of one more than the unknown." Good notation simplifies the expression of common ideas. It also leads to greater abstraction: we start thinking of matrices and vectors as *objects in themselves*, rather than as arrays of numbers. We also focus on the relationships between objects, and the rules governing their behavior (like $A(B + C) = AB + AC$), rather than the objects themselves.

Then it is an easy step to say "*A vector space is a set with the following operations and rules...*" and "*A linear transformation is a function $A\colon V \to W$ for vector spaces V and W where ...*" This is the beginning of an abstract theory. But when you create an abstract theory, you should always look back to the concrete questions that led to the theory.

5.3   Converting one kind of mathematics into another

It is easy to develop the idea that discrete and continuous mathematics are just two kinds of mathematics that do not connect. Deep developments in mathematics often come when we see how to connect things that previously looked unrelated. Here are some examples of going from continuous to discrete:

- *Bifurcations in differential equations*: The stability of an equilibrium point of a differential equation $d\mathbf{x}/dt = \mathbf{f}(\mathbf{x}, a)$ changes as a parameter a goes through a critical value. While the differential equation is a continuous thing in multiple ways, we can categorize points as equilibrium points $\mathbf{f}(\mathbf{x}, a) = 0$ and determine if they are stable or unstable. One of the next questions is: *When does the stability of an equilibrium point change?* Then we might ask: *What kinds of changes can occur in the structure of a differential equation?* Since we are looking at discrete outcomes, we can start asking questions that look more like discrete mathematics than continuous mathematics.
- *Reidemeister moves*: Reidemeister moves are discrete moves on a "knot diagram" (a special kind of network) that represents deformations of a continuous closed curve in three dimensions. Again, the emphasis is on when the "structure" of the knot (or knot diagram) changes significantly. See also Exercise 5.2 at the end of the chapter.

There are also examples of going from discrete to continuous:

- *Analytical number theory*: The most famous difficult theorem of number theory is the prime number theorem: if $\pi(n)$ is the number of prime numbers $\leq n$, then

$$\lim_{n \to \infty} \frac{\pi(n)}{(n/\ln n)} = 1. \tag{5.1}$$

  This was conjectured by C.F. Gauss around 1793 (unpublished), but only proved in 1896 independently by Jacques Hadamard and Charles Jean de la Vallée-Poussin. The methods of proof used by Hadamard and de la Vallée-Poussin were based on complex analysis, using some deep ideas of Bernhard Riemann.
- *Asymptotics of combinatorics*: Often combinatorial counting problems, like finding the number of unlabeled trees on n vertices, are hard to determine exactly with a simple formula. Instead, asymptotic estimates can be obtained via generating functions (see Section 4.8) and similar tools. Again, complex analysis is a powerful tool.

Other examples of converting one kind of mathematics into another include:

- Reducing geometric problems about constructing points with compass and straight-edge to questions about solving polynomial equations using only the usual arithmetic operations and square roots. This is part of Galois theory.
- Using measure theory to help answer questions about the long-time behavior of "chaotic" dynamical systems. So-called "strange attractors" have a fractal behavior, and so are very hard to describe explicitly. However, ergodic theory uses measure theory to describe these sets and the dynamics on them.
- Algebraic topology creates algebraic objects such as groups to describe (in a suitable way) the "shape" of objects that is invariant under continuous transformations with continuous inverses.
- The Perron–Frobenius theory of matrices with positive entries starts by showing that these must have positive eigenvalues. To do that, we can start by creating a function

$$\psi(\mathbf{x}) = \frac{A\mathbf{x}}{\sum_{j=1}^{n}(A\mathbf{x})_j}$$

which is a continuous function taking $\{\, \mathbf{x} \mid x_i \geq 0 \text{ for all } i,\ \sum_{i=1}^{n} x_i = 1 \,\}$ into itself. (The crucial point is that the denominator cannot be zero or negative for non-zero vectors with non-negative entries.) The fixed point (which exists by means of Brouwer's theorem from nonlinear analysis) is the desired eigenvector.

- Questions about existence of solutions of partial differential equations are often answered by means of functional analysis and/or Fourier transforms.

5.4   What is an interesting question?

Part of what makes an interesting answer is an interesting question. Finding interesting questions that mathematics can (potentially) answer is part of what makes a good mathematician. Consider, for example, counting primes less than or equal to a given number x. In fact, we can write

$$\pi(x) = |\{\, p \mid p \text{ prime and } 1 \leq p \leq x \,\}|\,.$$

Of course, given a real number x, we could in principle check every positive integer less or equal to x to see if it is prime, and return the number of primes found. We would have an algorithm for determining how many numbers are prime and $\leq x$. With this we could answer the question "*How many primes are* $\leq 32\,522$?" or "*How many primes are* $\leq 2\,375\,211$?"[1] Neither of these particular questions seem particularly interesting. If the values had some interesting regularity

---

[1] There are 3489 primes $\leq 32\,522$ and 174 593 primes $\leq 2\,375\,211$.

we might be on to something, but $\pi(x)$ has a pseudo-random character making specific values hard to predict.

More interesting might be *"How quickly does $\pi(x)$ grow as x increases?"* If we focus less on the particular values, and look for something less precise we have a more interesting question for which we might seek an answer. Here is an answer, which is known as the *prime number theorem*:

Theorem 5.1 (The prime number theorem). *If $\pi(x) = |\{\, p \mid p \text{ prime and}$ $1 \leq p \leq x \,\}|$, then*

$$\lim_{x \to \infty} \frac{\pi(x)}{(x/\ln(x))} = 1.$$

This only gives a rough idea about how many primes there are $\leq x$, but it brings understanding that the answer to *"How many primes are $\leq 2\,375\,211$?"* does not.

Relating apparently unrelated mathematical objects often leads to interesting objects and relationships as well as interesting questions. A case in point is relating $\pi(x)$ to the Riemann-ζ function:

$$\zeta(s) = \sum_{n=1}^{\infty} \frac{1}{n^s} \qquad \text{for Re}(s) > 1.$$

The connection with prime numbers is given by

$$\zeta(s) = \prod_{p \text{ prime}} \left(1 - p^{-s}\right)^{-1}.$$

See Exercise 5 at the end of this chapter. Now let us pose the question: *"How can we relate $\pi(x)$ to $\zeta(s)$?"*

This question about relating $\pi(x)$ and $\zeta(s)$ is not as ambitious a question as asking *"How are the primes distributed?"* But what would an answer to *"How are the primes distributed?"* look like? Relating $\pi(x)$ and $\zeta(s)$ would potentially lead to a partial answer to *"How are the primes distributed?"*, but should have a much clearer answer.

Good questions guide us in choosing what to investigate next. They are not necessarily the questions we *really* want to answer, but light the way towards answers to those more profound questions. Sometimes we focus on simpler problems than the ones we really want to solve. Hopefully the answers to those easier questions brings some enlightenment regarding the more complex questions. Sometimes we look at analogous questions in related but different parts of mathematics. Be prepared to look for counterexamples to your conjectures, for both the simpler and more complex problems.

5.5 Exercises

(1) There are four main models of computation: Turing machines, recursive functions, Church's λ-calculus, and Markov algorithms. Look these up, and outline a program for showing that they are all equivalent, in a suitable sense.

(2) Knot theory is about families of continuous functions $\mathbf{x}\colon [0,1] \to \mathbb{R}^3$ where $\mathbf{x}(0) = \mathbf{x}(1)$. Yet, almost all of the work on knots uses knot diagrams: planar graphs with nodes (of degree four) marked to show which thread goes "under" and which thread goes "over". Explain how it is that the uncountably infinite set of possible manipulations of a knot can be reduced to a finite number of Reidemeister moves on a knot diagram.

(3) Starting in the 1920's Stefan Banach and a group of mathematicians in Lwów, Poland started working on a theory of infinite-dimensional vector spaces. Since they needed to work on infinite sums, they needed notions of convergence and distance. Give a "rational reconstruction" of how the basic concepts of norms, Banach spaces and linear operators might be developed from this starting point. That is, give a presentation of these concepts in a way that is logical and *could have been* the way these concepts developed historically.

(4) Dirichlet proposed a minimization principle for solving the Poisson equation: the solution u of

$$\frac{\partial^2 u}{\partial x^2} + \frac{\partial^2 u}{\partial y^2} = f(x,y) \qquad \text{in a region } \Omega$$

with the boundary conditions $u(x,y) = g(x,y)$ for all (x,y) in the boundary, minimizes the functional

$$\int_\Omega \left[\frac{1}{2}\left(\frac{\partial u}{\partial x}\right)^2 + \frac{1}{2}\left(\frac{\partial u}{\partial y}\right)^2 + f(x,y)\,u(x,y) \right] dx\,dy$$

over all $u(x,y)$ subject to the same boundary conditions. It was soon criticized because there was no justification for the existence of a minimizer for the functional. What mathematical steps were needed to justify existence of a minimizer?

(5) The Riemann zeta (ζ) function is the function of a complex variable s where

$$\zeta(s) = \sum_{n=1}^{\infty} \frac{1}{n^s} \qquad \text{for } \mathrm{Re}(s) > 1.$$

Show that for $\mathrm{Re}(s) > 1$,

$$\zeta(s) = \prod_{p \text{ prime}} \left(1 - p^{-s}\right)^{-1},$$

where the product is over all primes p. Relate $\zeta(s)$ to $\pi(x)$, the number of primes $\leq x$. [**Hint:** If $f: [1, \infty) \to \mathbb{R}$ is continuous, then $\int_1^\infty f(x)\, d\pi(x) = \sum_{p \text{ prime}} f(p)$ where the integral is a Riemann-Stieltjes integral. We can also do integration by parts: $\int_a^b f(x)\, dg(x) = f(x)g(x)|_{x=a}^{x=b} - \int_a^b g(x)\, df(x)$ provided f is continuously differentiable and g has bounded variation. Note that "$df(x)$" is the same as "$f'(x)\, dx$".]

(6) Show that for $\text{Re}(s) > 1$,

$$\frac{1}{\zeta(s)} = \prod_{p \text{ prime}} \left(1 - p^{-s}\right) = \sum_{n=1}^\infty \frac{\mu(n)}{n^s}$$

where $\mu(1) = 1$, $\mu(n) = (-1)^k$ where n is the product of exactly k distinct primes, and $\mu(n) = 0$ if n has a factor of the form p^r with $r > 1$.

(7) Suppose X is some kind of continuous "shape" or "space" containing a designated point x_0. A *loop* in X is a continuous function $x: [0, 1] \to X$ where $x(0) = x(1) = x_0$. We say two loops x and y are *homotopic* ($x \sim y$) if we can deform one into the other continuously: that is, there is a continuous function $h: [0, 1] \times [0, 1] \to X$ where $h(s, 0) = x(s)$ and $h(s, 1) = y(s)$ for all $0 \leq s \leq 1$, and $h(0, t) = x_0$ for all $0 \leq t \leq 1$. Show that "$\sim$" is an equivalence relation. Then show that we can define an operation "$*$" on the equivalence classes by $[z]_\sim = [x]_\sim * [y]_\sim$ where $z(s) = x(2s)$ for $0 \leq s \leq \frac{1}{2}$ and $z(s) = y(2s - 1)$ for $\frac{1}{2} \leq s \leq 1$. Show that this operation is associative and has an identity and has inverses. This algebraic operation forms a group called the *fundamental group*, denoted $\pi_1(X, x_0)$.

(8) Look up some famous conjectures, and the mathematics that they led to:

    (a) Fermat's last "theorem".
    (b) Mordell's conjecture.
    (c) the Bieberbach conjecture.
    (d) the Goldbach conjecture.
    (e) the "$P = NP$" conjecture.

(9) Factoring polynomials has been part of mathematics at least since the 17th century, and part of high-school mathematics. A polynomial that cannot be factored is called *irreducible*. But which polynomials are irreducible depends on what coefficients are allowed. The Fundamental Theorem of Algebra says that any non-constant real or complex polynomial of degree n can be factored into a product of n linear factors with complex coefficients. Whether a polynomial is irreducible using only rational or only integer coefficients is an interesting question. See if you can relate factorization of

polynomials over integers to factorization over integers modulo p for a prime number p.

(10) Propose a definition for pseudo-convex sets on the unit sphere $S^2 = \{ \mathbf{x} \in \mathbb{R}^3 \mid \mathbf{x}^T \mathbf{x} = 1 \}$. Using this definition, show that the intersection of a collection of pseudo-convex sets is also a pseudo-convex set in S^2, or give a counterexample to show that this is not true. Define the pseudo-convex hull of a subset A of S^2 as the intersection of all pseudo-convex sets containing A. Give some examples of pseudo-convex hulls for your definition.

Bibliography

Bollobás, B. (1979). *Graph theory: An introductory course, Graduate Texts in Mathematics*, Vol. 63 (Springer-Verlag), ISBN 0-387-90399-2.

Flajolet, P. and Sedgewick, R. (2009). *Analytic combinatorics* (Cambridge University Press, Cambridge), ISBN 978-0-521-89806-5, doi:10.1017/CBO9780511801655.

Frege, G. (1879). *Begriffsschrift, eine der arithmetischen nachgebildete Formelsprache des reinen Denkens* (Verlag von Louis Nebert, Halle).

Gersting, J. L. (2007). *Mathematical Sctructures for Computer Science*, sixth edn. (W.H. Freeman & Co.), ISBN 978-0-7167-684-X.

Hardy, G., Littlewood, J., and Pólya, G. (1952). *Inequalities*, 2nd edn. (Cambridge University Press), ISBN 0521-052068.

Hardy, G. H., Wright, E. M., and Wiles, A. (2008). *An introduction to the theory of numbers*, sixth edn. (Oxford University Press, New York), ISBN 978-0199219865.

Lax, P. D. (2009). Rethinking the Lebesgue integral, *Amer. Math. Monthly* **116**, 10, pp. 863–881, doi:10.4169/000298909X476998.

Pommersheim, J., Marks, T., and Flapan, E. (2010). *Number Theory: A Lively Introduction with Proofs, Applications, and Stories*, 1st edn. (Wiley), ISBN 978-0470424131.

Whitehead, A. N. and Russell, B. (1925–1927). *Principia Mathematica* (Cambridge University Press), not for the faint of heart!

Index